GREAT
BREAKTHROUGHS IN
PHYSICS

GREAT BREAKTHROUGHS IN
PHYSICS

HOW THE STUDY OF MATTER AND ITS MOTION CHANGED THE WORLD

ROBERT SNEDDEN

SIRIUS

SIRIUS

This edition published in 2020 by Sirius Publishing, a division of
Arcturus Publishing Limited,
26/27 Bickels Yard, 151–153 Bermondsey Street,
London SE1 3HA

ISBN: 978-1-83940-685-0
AD007208UK

Printed in China

Contents

Introduction

SUPERSTITION TO SUPERTHEORIES

> *'In physics, you don't have to go around making trouble*
> *for yourself – nature does it for you.'*

American theoretical physicist Frank Wilczek (1951–)

In many ways it is the purpose of science to explain and interpret reality and to find the underlying rules and order that drive natural processes. Physics can be defined as the science of matter and energy and the ways in which the two interact with each other. In other words, it is the science that underpins absolutely everything, or as Ernest Rutherford (1871–1937) is alleged to have remarked, 'all science is either physics or stamp collecting'.

Physics is concerned with acquiring knowledge and understanding the world around us. In order to thrive our ancestors had to learn how the world works through a combination of observation and experiment. Humans became scientists through a pressing need to survive in their environment. Toolmaking, for example, involved selecting those materials that had the necessary characteristics for the job at hand – some rocks would take a sharp edge, others would not. The first toolmakers had to experiment and discover 'what happens if...?' The need to gain practical knowledge, and to pass that knowledge on to others, sowed the first seeds of science.

People learned by trial and error which stones were best for shaping tools.

The thinkers of Ancient Greece preferred to discuss their ideas rather than test them by experiment.

Similarly, early humans would have realized that many natural processes have a predictable regularity. The sun will rise each morning, spring will follow winter and a thrown rock will always fall to the ground. It is understandable that in attempting to explain this natural order people would turn to religion and magic.

Perhaps the first thinker known to us who embraced a recognizably scientific way of looking at things was Thales of Miletus (*c.*624–*c.*547 BC). Thales, one of the semi-legendary Seven Sages of Ancient Greece, believed that, contrary to the prevailing belief in the power of the supernatural, all phenomena could be explained in natural, rational terms. His view was that to understand the world, it was necessary to know its nature, or 'physis' (from which we get 'physics'). An essential part of the approach taken by Thales and his fellow Greek philosophers was that their ideas were subject to scrutiny in open, critical debates and that all theories and explanations could be challenged. It is a way of doing things that still informs scientific enquiry to this day. Where the proto-scientists of Greece differed from present day science, was in believing it unnecessary to test their hypotheses through experiment; a well-reasoned self-consistent argument was thought sufficient.

Aristotle (384–22 BC) promoted the concept of natural laws for physical phenomena. His range of interests was extraordinary and, in addition to physics, included philosophy, logic, astronomy, biology, psychology, economics, poetry and drama. Aristotle's ideas effectively dominated western science and philosophy for nearly 2,000 years, becoming enormously popular in Europe with the scientific developments of the Middle Ages until they were challenged by Galileo at the beginning of the 17th century.

Over time, the view grew that natural events were guided by natural laws rather than supernatural interventions, laws which could be uncovered and understood. One of the first to make reference to the concept of a law of nature in something approaching the modern sense, was the Franciscan friar and scholar Roger Bacon (c.1214–92). Bacon is said to have admonished people to 'Cease to be ruled by dogmas and authorities; look at the world!' Bacon's independent thinking brought him into conflict with the Catholic church, which embraced religion and the authority of Aristotle on matters of science. He was imprisoned for 15 years for heresy, but he had prepared the way for a scientific investigation of the natural world that would be rooted in experiment and reason.

Another of Bacon's dictums was that 'Mathematics is the door and the key to the sciences' and that is beyond doubt true of physics, which is largely concerned with things that can be measured and quantified. The most reliable way to discern meaningful patterns in the data gathered by experiment and observation is through mathematical analysis. By Bacon's time mathematicians were developing the powerful new tool of algebra, which allowed unknown quantities to be expressed by symbols and gave scientists the means to explore general relationships in ways that just hadn't been possible before. By the 16th and 17th centuries, physicists such as Galileo Galilei (1564–1642) and Isaac Newton (1642–1727), both incidentally professors of mathematics, were opening up a new era of physics married to the precision of mathematics.

Roger Bacon was one of the first to champion observation and experiment and to challenge the accepted authority of Aristotle.

Copernicus delayed publication of On the Revolutions of the Celestial Spheres *for several years, fearing the reception it might get.*

Prior to Galileo and Newton, a literally earth-moving revolution took place in the scientific view in 1543 with the publication of Nicolaus Copernicus' (1473–1543) rejection of the prevailing earth-centred view of the universe in favour of a heliocentric perspective in which the earth moved round the sun. By removing humanity from its God-given place at the centre of creation, Copernicus had set up an inevitable confrontation with the Catholic church. Galileo was summoned to appear before the Inquisition in 1633, on a charge of heresy, for supporting Copernicus' ideas. Coerced by the threat of torture into renouncing the idea that the earth moves around the sun a defiant Galileo is said to have muttered '*Eppur si muove*' ('And yet it moves').

Mathematics, as Bacon had predicted, did indeed become key to opening up the physics of the universe. Johannes Kepler (1571–1630), using the best observational data available at the time, demonstrated in 1609 that the planets *did* move around the sun, but, contrary to belief and expectation, their orbital paths were ellipses rather than circles. This discovery underpinned the importance of data over belief. Kepler hadn't expected to discover the planets moving in ellipses, but there was no denying the data and the mathematics.

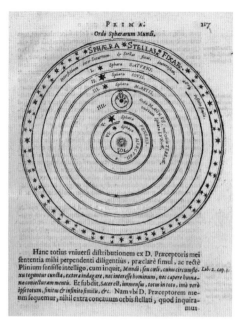

Before the data proved the idea wrong, it was believed the planets moved in perfectly circular orbits.

It fell to Newton to explain the cause of Kepler's ellipses. In 1684, astronomer Edmond Halley (1656–1742) asked Newton how a planet would move if it was attracted towards the sun by a force that weakened in proportion to the inverse square of its distance. Newton, having apparently worked out the answer years previously, is said to have given Halley the answer immediately: an ellipse. Halley persuaded Newton to publish his calculations. The result, in 1687, was a contender for the most influential book in the history of science. Newton's *Philosophiae Naturalis Principia Mathematica* (*Mathematical Principles of Natural Philosophy*), or simply the *Principia*, set out Newton's definitions of force and mass, and his three laws of motion. In Part 3 of the book, in his *System of the World*, Newton proposed that gravity was a universal force, acting between all objects, with a force proportional to the product of their masses and the inverse square of the distance between them and so explained the motion of the planets unveiled by Kepler.

Following Newton's unparalleled insights scientists began to entertain the notion that it was possible to quantify and understand everything. Newton had provided a framework around which a coherent rational worldview might be built.

In 1814, Pierre-Simon de Laplace (1749–1827) published an essay, *Essai philosophique sur les probabilités* ('A philosophical essay on probabilities'), on the deterministic universe. He imagined the existence of a super-intelligent being, which became known as Laplace's Demon, that could, at one instant, know the positions and velocities of all the objects in the universe, and the forces acting upon them, and with this knowledge have the ability to calculate their positions and

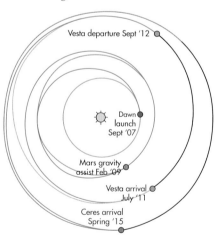

Newton's laws are accurate enough to plot the flightpaths of spacecraft.

Pierre-Simon de Laplace imagined an all-knowing demon that could predict the future shape of the universe.

velocities for all future times. His 'demon' depended for his foreknowledge on a predictably mechanistic view in which all future events are a consequence of past conditions. It was a view that appeared to embrace predestination and to deny any possibility of free will.

Up until around the end of the 19th century, physics appeared to be almost entirely concerned with ways of refining the mechanistic view of the universe. The invention of the steam engine, powerhouse of the Industrial Revolution in the 18th century, spurred the development of thermodynamics as scientists and engineers looked for ways to squeeze every last drop of efficiency from their machines. The nature of heat, energy and work was investigated more thoroughly than it ever had been before.

These enquiries gave rise to the powerful idea of entropy, the measure of disorder in a system and the idea, enshrined in the second law of thermodynamics, that entropy always tends to increase, along with the ideas of equilibrium and irreversibility. Together, these notions pointed towards the ultimate running down of the universe, in what cosmologist Sir Arthur Eddington (1882–1944) dubbed a final state of 'heat death'.

Eddington had a part to play in promulgating the ideas of one of the greatest physicists who ever lived – Albert Einstein (1879–1955). Before Einstein, space and time were simply the backdrop upon which the events that interested physicists took place. Relativity changed that. Around 1864, James Clerk Maxwell (1831–79) had predicted the speed of light from the fundamental constants that underpin electromagnetism. In 1905, Einstein saw that the speed of light would have to remain a constant for all observers and from this deceptively simple premise drew incontrovertible conclusions that involved shrinking and stretching space and time themselves. As Einstein's former teacher put it: 'Henceforth space by itself, and time by itself, are doomed to fade away into mere shadows, and only a kind of union of the two will preserve an independent reality.' Einstein went on to show that gravity, rather than being the force envisaged by Newton, was actually the result of spacetime being distorted by the presence of matter embedded in it. As American physicist John Archibald Wheeler succinctly described it: 'Spacetime tells matter how to move; matter tells spacetime how to curve.'

FIG. II.
Art 119

Lines of Force and Equipotential Surfaces.

$A = 20$ $B = -5$ P, *Point of Equilibrium.* $AP = 2AB$
Q, *Spherical surface of Zero potential.*
M. *Point of Maximum force along the axis*
The dotted line is the Line of Force $\Psi = 0.1$ *thus* ————

James Clerk Maxwell mapped out lines of force in his groundbreaking Treatise on Electricity and Magnetism *(1873).*

At the beginning of the 20th century there came a profound change in the way we look at the universe, a scientific watershed that divided physics into the classical physics that came before and the quantum physics that came after. Some physicists even regard Einstein's relativity theories as belonging in the classical camp rather than marking a serious departure from it. But Einstein had a hand in the birth of the quantum era as well.

In 1905, Einstein published his explanation of the photoelectric effect, the release of electrons from a metal exposed to electromagnetic radiation. To do so, he used a notion that had been introduced into physics five years earlier by Max Planck (1858–1947). This was the idea that energy, rather than being a continuously variable quantity as had always been assumed, actually came in discrete packets, called quanta.

The quantum of energy was the seed that became a revolution in physics. As scientists such as Niels Bohr (1885–1962), Werner Heisenberg (1901–76) and Erwin Schrödinger (1887–1961) began to explore its implications they were constructing a view of the universe where nothing was certain, where light could be simultaneously both a wave *and* a particle, where the exact same experiment could have different outcomes and where the properties of an object have no real meaning until you measure them. As far as quantum physics is concerned there is no 'real' world out there, only an amorphous sea of probability and potential.

One of the biggest challenges for physics is finding a way to reconcile the fundamental forces of the quantum realm with the gravity and warped spacetime of Einstein's universe in a so-called 'theory of everything'. It is a goal that has so far proved to be elusive, evading the best efforts of thinkers such as Stephen Hawking (1942–2018) and Einstein himself. Hawking came to believe that an overarching supertheory would remain forever out of reach because our human perceptions of reality were always going to be incomplete. Yet Hawking was undismayed by this, declaring: 'We have this one life to appreciate the grand design of the universe, and for that, I am extremely grateful.'

In this book we'll take a look at just some of the discoveries that physics has made on the way to understanding that grand design.

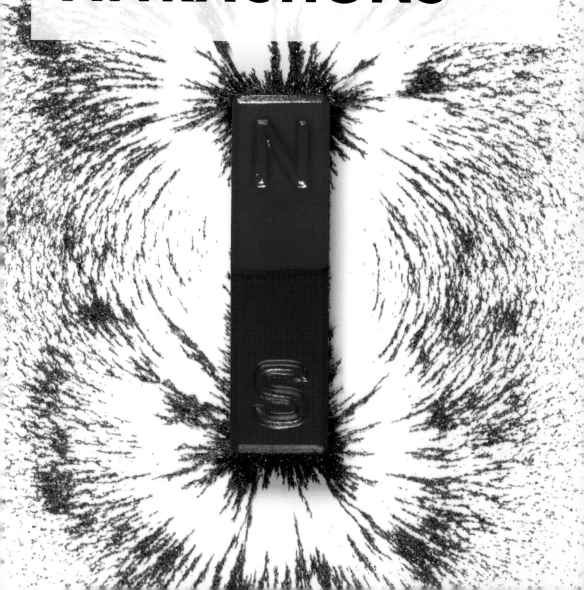

Chapter 1
PHYSICAL ATTRACTIONS

Physical Attractions

TIMELINE OF DEVELOPMENTS IN MAGNETISM

Timeline	
6TH CENTURY BC	Greek philosopher Thales describes the ability of lodestone to attract iron.
4TH CENTURY BC	The Chinese use lodestones as fortune-telling aids.
1088	Chinese scientist Shen Kua gives one of the first descriptions of lodestone's ability to magnetize iron.
1190	Alexander Neckam writes about sailors from the East using magnetized needles for navigation.
1269	Petrus Peregrinus writes the first complete account of magnetism in the *Epistola de magnete.* He is the first to describe the concept of polarity, naming the magnetic north and south poles.
14TH CENTURY	The nautical compass is in common use.
1544	Georg Hartmann records his observations of magnetic declination, the discrepancy between geographic north and magnetic north.
1600	William Gilbert publishes *De Magnete*, perhaps the first ever work on experimental physics. Gilbert believed that the earth itself was a giant lodestone, and its magnetic field was responsible for the earth's rotation and keeping the moon in orbit.
1698	Edmond Halley sets out on HMS *Paramore* on an expedition to make the first magnetic chart of the Atlantic Ocean.

The phenomenon of magnetism has been known about for millennia. Around the 4th century BC the Chinese were using what they called 'south pointers', not to find their way, but to ensure that their houses were pointed in the most auspicious direction. These fortune-telling aids used lodestones, a form of magnetite, a naturally occurring magnetic mineral. The oldest description of lodestones in Europe dates back to the Greek philosopher-scientist Thales, who, around 600 BC, noted their ability to attract iron.

Thales also observed that rubbing a piece of amber, a fossilized resin, with a fur cloth gave it the power to attract small objects such as feathers or bits of straw – a phenomenon that was actually due to static electricity (see pages 74–6). The deep relationship between the attractive powers of lodestone and amber was something that would not be brought to light for over 2,000 years (see page 75).

The magnetic attraction of lodestones was a mystery that was hard to explain. Thales thought that magnets had a soul, that the ability to initiate motion in something else was a sign of life. The Greek thinker Democritus argued that the lodestone's power resulted from its tendency to emit particles or 'effluvia' that carved out a void in space that other objects would rush to fill. Lucretius, a follower of Democritus, tried to address the obvious problem with this explanation. Why doesn't the lodestone attract other materials besides iron? He argues that gold, for example, is too heavy, so it doesn't move, and lighter substances such as wood let the effluvia pass right through, so it doesn't get reflected and sweep away the air between the two substances.

Pliny (AD 23–79) wrote: 'What phenomenon is more astonishing? Where has nature shown greater audacity? For iron, the tamer of all substances... leaps to meet the magnet.' Pliny recounts the story of a shepherd called Magnes in northern Greece around 800 BC who was astonished to find that his iron-tipped staff and the nails in his shoes were being drawn to a particular rock. Magnes gave his name to the region of Greece in which he lived, Magnesia, and hence led to the phenomenon he experienced being called 'magnetism'.

The early Christian philosopher St Augustine (AD 354–430) was astonished the first time he saw a magnet lift a chain of rings, and by the power of a

The mythical Magnes the shepherd who was said to have discovered the magnetic properties of lodestone.

magnet held beneath a silver plate to move a piece of iron around on the top of the plate. The fact that the lodestone would not move straw, whereas amber did, was a puzzle to him.

That iron could itself be magnetized by stroking it with a lodestone was an important discovery that had doubtless been known about for centuries. One of the first references to this ability was by the Chinese scientist and mathematician Shen Kua (1031–95) in 1088; the first European mention was by English theologian Alexander Neckam (1157–1217) in 1190. Neckam's work *De Naturis Rerum* (*On the Nature of Things*), mentions that sailors from the East used magnetized needles for navigation, the first textual reference in Europe of the use of compasses. By the 14th century the nautical compass was in regular use by the English navy. Christopher Columbus carried one on his voyage of discovery to the Americas in 1492.

St Augustine was perplexed by the power of magnetism.

PETRUS PEREGRINUS

French engineer Pierre de Maricourt, known as Petrus Peregrinus, who lived during the 13th century, was one of the few to conduct systematic experiments on magnetism during the medieval period. His *Epistola de magnete* (Letter on the magnet), completed in 1269, was the first comprehensive account of magnetism. He described how to make a compass, instructing the reader that 'By this very instrument, you may direct your course

Medieval philosophers, such as Petrus Peregrinus, were fascinated by the properties of magnets.

to any cities and islands, to whatever other places, wherever you wish to by land or sea, with the longitude and latitude of these places always known to you.'

His experiments with magnets, including methods of determining magnetic polarity and the effects of magnetic attraction and repulsion, involved placing an iron needle on a spherical magnet and drawing a line along the direction of alignment. By moving the needle to different positions and repeating the process Peregrinus discovered that the pattern of lines converged at two points opposite each other on the sphere. He seems to have been the first to describe these points as *polus*, defining the concept of polarity for the first time and naming the north and south poles of the magnet. He observed the ability of a strong magnet to reverse the polarity of a weaker one and was also the first to formulate the idea that opposite poles attract whereas like poles repel. He also noted that when a magnet was broken in two each half had its own north and south poles.

DECLINATION AND INCLINATION

At some point it became apparent to compass-steering navigators that the direction their compass needles were pointing in was not exactly towards geographic, or true, north. Just when it was realized that allowances had to be made for this discrepancy is not clear. Christopher

Columbus recorded noticing variations in his compass readings during his 1492 voyage and it is most likely that earlier explorers had noticed the discrepancy too. This difference in the direction of magnetic north from that of geographic north varies across the surface of the earth and is referred to as the magnetic declination. A letter of 1544, written by Georg Hartmann, a vicar of Nürnberg, recorded his observation that the declination at Nürnberg in 1510 was 10°E, whereas in Rome it was 6°E.

Hartmann's letter also mentioned his discovery that a magnetized needle didn't only swing north, if it was free to move vertically, it also pointed downwards. Unfortunately, his letter was filed away and ignored and he didn't get the credit for this. In 1581, Robert Norman, a scientific instrument maker

By the 14th century compasses were becoming indispensable aids to shipboard navigators.

in London, published a work that described his independent discovery, in 1576, of the phenomenon of magnetic dip, or inclination.

Among other devices, Norman made compasses for use on ocean-going ships. These compasses were made by magnetizing an iron needle using a lodestone and then balancing it on a support. Norman noticed that when he did this the north-seeking end of the compass needle always tilted slightly downwards, forcing him to add a small counterbalancing weight to the other end. Norman set out to investigate the effect more thoroughly. He experimented by mounting his needle on a small ball of cork, which he weighted to give it neutral buoyancy, so that it would neither sink to the bottom nor float to the top, and submerging it in water, to cancel out the effect of gravity. The magnetized needle still pointed north but now tilted downwards as well. The force that attracted the needle northwards was also drawing it towards the earth. The mystified Norman never did find out what caused the tilt of the magnetized needle, but he had made the first tentative steps towards the discovery of the magnetic field.

DE MAGNETE

Robert Norman died around 1600, the year in which William Gilbert (1544–1603) published a work that made substantial advances in our knowledge of magnetism. It is next to impossible to assign a starting point for the beginning of the modern scientific era. There is a good case to be made for the revolution brought about by Copernicus in 1543 when he shifted the earth from its place at the centre of the universe, or for the mechanical and astronomical discoveries of Galileo at the beginning of the 17th century. Another contender is the publication of one of the first great works of physics and experimental science: *De Magnete, Magneticisque Corporibus, et de Magno Magnete Tellure* (*On the Magnet, Magnetic Bodies, and the Great Magnet of the Earth*). Gilbert's six-volume magnum opus provided an account of his systematic

William Gilbert demonstrates some of his findings to Queen Elizabeth I.

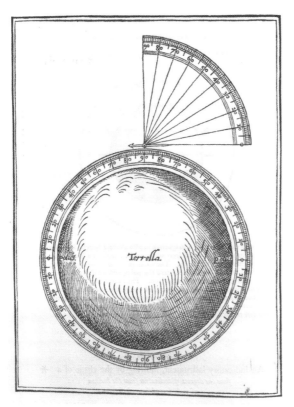

An illustration from De Magnete *showing the measurement of declination using a* terrella.

researches into magnetism that heralded a new age of scientific experiment and enquiry.

Gilbert's investigations into magnetism took place between around 1580 and 1600, at a time when he was also enjoying a very successful career as a physician (he became President of the Royal College of Physicians in 1599 and was appointed physician to Queen Elizabeth I in 1601).

Prior to the publication of *De Magnete*, which has been described as the first ever work of experimental physics, the idea of a science textbook was practically unknown. Gilbert's careful use of experiments marked him out as one of the forerunners of the modern scientific method. As he wrote in the preface to *De Magnete*: 'In the discovery of secret things and in the investigation of hidden causes, stronger reasons are obtained from sure experiments and demonstrated argument'.

Although he did acknowledge a debt to Peregrinus, Gilbert found the existing literature on magnetism to be somewhat lacking, and set out to make his own investigations of the phenomenon. He carried out experiments using a spherical lodestone, which he called a *terrella*, or 'little earth', as a model for the earth. (Present-day scientists have employed terrellas inside vacuum chambers to, for example, mimic the effect of the earth's magnetic field on cosmic ray particles and the solar wind.)

By moving a small compass needle over the surface of the *terrella*, Gilbert was able to observe how the direction of the needle changed from place to place. He demonstrated that his *terrella* had a north and south pole and also that the needle dipped the nearer it was moved towards a pole, just as the needle on a mariner's compass dipped on the surface of the earth. Speculation as to the source of the magnetism that attracted the compass needle included the Pole Star and the existence of an as yet uncharted magnetic island somewhere near the North Pole, but Gilbert made the bold assertion that the whole planet earth was a giant lodestone. He defined the north magnetic pole as the place where a magnetized needle will point vertically downwards.

Gilbert knew that magnetic poles can attract or repel, depending on their polarity. In addition, however, he noticed that ordinary iron is *always* attracted to a magnet, never repelled by it. He surmised that, when brought close to a permanent magnet the iron became a temporary magnet of a polarity that would be attracted to the permanent magnet. In other words, an iron bar brought near the south pole of a magnet temporarily becomes a magnetic north pole. Because magnetic poles are never found singly, the other end of the iron temporarily becomes a south pole, and is able to attract more iron to it. It's easy to see this by picking up paper clips with a magnet – it's obvious that not all the paper clips are in direct contact with the magnet. Gilbert confirmed his guess by hanging two parallel iron rods above the pole of a *terrella*. When he did so the rods repelled each other, each having become a temporary magnet with the same polarities.

In Book 6 of *De Magnete*, Gilbert put forward his belief that the earth rotates on its axis and suggested that magnetism was the cause of this, writing that: 'Were not the earth to revolve with diurnal rotation, the sun... would scorch the earth, reduce it to powder... In other parts all would be horror, and all things frozen stiff with intense cold... the earth seeks and seeks the sun again... by her wondrous magnetical energy.' Although he didn't fully commit himself to the Copernican notion of a few decades earlier that the earth orbits the sun, in a later work Gilbert went on to say that magnetic attraction was responsible for holding the moon in orbit around the earth, and for the moon's influence on the tides.

Gilbert's experiments also demonstrated the difference between electrical and magnetic phenomena. He conducted experiments into electrical phenomena using a sensitive 'versorium', a piece of equipment that he invented himself (see page 75). At all times, Gilbert took care to describe his equipment and experimental set-ups accurately so that others could replicate them and confirm his results and findings for themselves. It was a significant breakthrough from the works of previous centuries which, as Gilbert himself described it, had 'never a proof from experiment, never a demonstration... Hence such philosophy bears no fruit... because few of the philosophers are themselves investigators, or have any first-hand acquaintance with things ...'

MAGNETIC MAPPING

Scientists struggled to come up with an explanation for the fact that the earth's magnetic north pole, and its 'true' or geographic north pole were not in the same place. Navigators plotting a course across the oceans relied on the magnetic compass and had to know how much of a discrepancy there was between the two. Complicating matters, in 1635 English mathematician Henry Gellibrand (1597–1637) published evidence that the difference between the magnetic and geographic poles was not constant but changed over time. It meant that navigators couldn't take compass bearings for granted as they became inaccurate over a period of decades and had to be recalibrated. Why this should be was a complete mystery. If the earth was indeed a magnet it was behaving like no other magnet known.

In 1692, physicist and astronomer Edmond Halley, more famous for the comet that is named after him, made an inventive suggestion. The earth, he claimed, was not a solid ball of rock but was formed from an outer shell and an inner core. Halley's proposal was that the earth actually had four magnetic poles, one pair at either end of the axis of the outer magnetic shell and the other pair on the axis of the inner magnetic core. Each sphere was independently magnetized, and rotated slowly with respect to the other, which accounted for the gradual 'drift' of the magnetic poles. As much later research would establish, he wasn't too far off the mark.

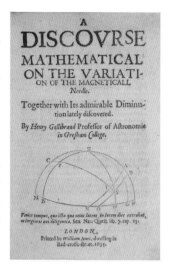

The title page of Gellibrand's book on the magnetic and geographic poles.

Edmond Halley, who was astronomer royal from 1720 to 1742.

Earthquake studies have shown that the earth does have a layered structure and that differences in the rate of spin between the solid inner core and the outer liquid core may be responsible for generating the earth's magnetic field.

Halley also carried out extensive work to map the magnetic variations across the earth's surface. He believed that these variations held the key to solving the problem for navigators of determining longitude – measuring their position east or west of a location. Halley was convinced that mapping east-west changes in magnetic variation could be used to chart changes in longitude. Without a reliable way of determining the longitude, navigators could never be quite sure where their ship was located on the ocean. Despite its

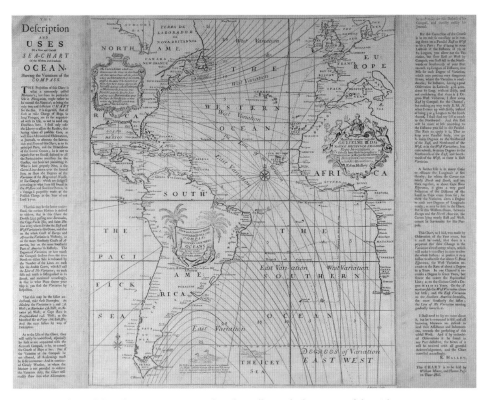

A magnetic chart of the Atlantic Ocean, mapped out by Halley at the beginning of the 18th century.

crucial importance for exploration and trade, a practical solution for working out longitude at sea had still not been found. The governments of sea-going nations were prepared to offer substantial rewards to anyone who could find an answer.

In 1698, Halley commanded a small ship called the *Paramore* on a voyage to map the magnetic field of the Atlantic Ocean. Surviving the perils of fog and icebergs, and an onboard falling out with the naval officer in charge of the ship, Halley created the first magnetic chart. Although it proved unsatisfactory for the purpose of determining longitude (because the variations fluctuated over time) Halley's chart was widely used throughout the 18th century. It was significant because it was the first to adopt isogonic lines (called 'Halleyan lines' at the time) to connect points of equal magnetic variation.

Carl Friedrich Gauss devised a method of measuring the intensity of the Earth's magnetic field.

Gauss' observatory in Göttingen.

Carl Friedrich Gauss (1777–1855) was a professor of mathematics at the German university of Göttingen. In 1828 he attended a conference in Berlin, where naturalist Alexander von Humboldt (1769–1859) showed Gauss his collection of magnetic instruments and encouraged him to study the phenomenon. Together with his assistant Wilhelm Weber (1804–91), Gauss devised an ingenious method of using an auxiliary magnet to measure not only the direction of the earth's magnetic force, but also its intensity. This made it possible to set up a global network of magnetic observatories, as it allowed every instrument to be calibrated locally, independently of any others.

The readings from the network of observatories could be combined using a precise mathematical method known as spherical harmonic analysis to build up an accurate map of the earth's magnetic field for the first time. Today's models of the magnetic field are derived largely from satellite data.

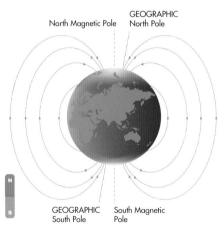

The Earth's magnetic field.

Chapter 2
CLASSICAL GAS

Classical Gas

TIMELINE OF GAS THEORIES

Timeline	
1609	Jan Baptist van Helmont discovers that heating coal in the absence of air produces 'wild gas' – now known as coal gas – a combination of methane, carbon monoxide and hydrogen.
1643	The barometer is invented by Evangelista Torricelli.
1653	Richard Towneley conducts an experiment with Henry Power and discovers that air pressure at the top of a hill is less than at the bottom.
1662	Robert Boyle publishes his finding that if the pressure is doubled, the volume of a gas is halved. This is now known as Boyle's law.
1783	Jean-François Pilâtre de Rozier makes the first hot air balloon flight over Paris in a balloon designed by the Montgolfier brothers.
1787	Alexandre Cesar Charles determines Charles' Law linking the volume of a gas to its temperature.
1801	Joseph Louis Gay-Lussac establishes that the pressure of a gas will vary according to its temperature when mass and volume remain constant.

Although Air was one of the four elements of Ancient Greece (along with Earth, Fire and Water) no real study was made of the properties of gases until the 17th century.

The word 'gas' was seemingly first suggested by the 17th century Flemish chemist Jan Baptist van Helmont (1580–1644), following his Dutch pronunciation of the Greek word 'chaos'. Until van Helmont's time all gases had been considered as varieties of air. Van Helmont was one of the first to describe how certain chemical reactions liberated substances that were air-like in their behaviour but which had properties that were distinct from those of atmospheric air. In an experiment, he burned 28 kilograms of charcoal and found that only

Jan Baptist van Helmont may have been the first to use the word 'gas'.

half a kilogram of ash remained, concluding that the rest had escaped as 'wild gas'. In 1609, Van Helmont described his discovery that heating coal in the absence of air produced a type of gas. Now commonly known as coal gas, this was in fact a mixture of methane, carbon monoxide and hydrogen. In the 18th and 19th centuries it was used to provide lighting for homes and factories.

MR TOWNELEY'S HYPOTHESIS

In 1661, Anglo-Irish philosopher, chemist and physicist Robert Boyle (1627–91) was contacted by Richard Towneley (1629–1707) a land owner in Lancashire and himself a keen student of astronomy and meteorology. Some years earlier, in 1653, Towneley, in company with his doctor friend Henry Power, had carried a barometer to the top of the 557-metre high Pendle Hill. The two investigators discovered that the air pressure at the top of the hill was less than that at the bottom of the hill. Furthermore, they established that the same quantity of air occupied a greater volume at lower pressure and a lesser volume at higher pressure.

Intrigued, Boyle recruited his assistant Robert Hooke (1635–1703), a talented scientist and inventor in his own right, to investigate what he dubbed 'Mr Towneley's hypothesis'. This was rather unfair to Power who had written up their findings in his as-yet-unpublished book *Experimental Philosophy*. Boyle, even having had sight of the manuscript, still insisted on giving sole credit to Towneley. Compounding the injustice, by the time Power did publish in 1663, Boyle's scientific status had ensured that it was his name that would be indelibly associated with the law that established the relationship between the pressure and the volume of a gas.

The barometer

Barometers soon found a place in homes as useful predictors of the weather.

The invention of the barometer is generally credited to Italian physicist and student of Galileo, Evangelista Torricelli (1608–47). In an experiment carried out around 1643, Torricelli filled a glass tube with mercury, sealed it at one end, and inverted it in a small basin which was also filled with mercury. He observed that, rather than run out into the basin, the column of mercury in the tube maintained a height of about 76cm. Torricelli correctly deduced that it was the pressure of the atmosphere pushing down on the mercury in the basin that caused the mercury in the tube to rise. Later experiments would show that the level of the mercury column varied with the changes in atmospheric pressure at different altitudes and in different weather conditions. French philosopher René Descartes (1596–1650) may have been the first person, probably in 1647, to attach a graduated scale to the mercury tube, making it easier to record changes day to day. Descartes, along with physicist Blaise Pascal (1623–62) wondered what would happen to the mercury column if the experiment was performed at the top of a mountain and, around 1648, Pascal's brother-in-law, Florin Perier, took a tube to the summit of the Puy de Dôme (an altitude of 1,465 metres) where he demonstrated that the mercury level was indeed several centimetres lower than the measure he obtained in the town of Clermont below. Torricelli's invention was soon seeing service as a piece of laboratory apparatus, as an instrument for measuring altitude, and monitoring and predicting the weather. The word *barometer*, which means an 'instrument for measuring weight', was first used by Robert Boyle in the 1660s.

Barometer readings demonstrated that the air pressure on hilltops was lower than at ground level.

BOYLE'S LAW

Boyle's reputation ensured that his name was attached to the relationship between pressure and volume in a gas.

Satisfied that Hooke's experiments and measurements had confirmed the findings of Power and Towneley, Boyle published the results in 1662. Boyle thought that air was made up of coiled particles that could be compressed, but which would spring back when the pressure was released. 'There is a spring, or elastical power, in the air we live in', he wrote. The experimental apparatus he described was simple but elegant, involving the use of J-shaped glass tubes sealed at one end. Mercury was poured into the open end of the tube, trapping a small volume of air in the sealed end of the tube. It had been established that atmospheric pressure would support 76 centimetres of mercury, so this gave Boyle a measure for gauging the pressure he was exerting on the trapped air. Boyle's experiments established a simple relationship:

if the pressure was doubled, at constant temperature, the volume of the gas was reduced by a half; if the pressure was tripled, volume was reduced to a third, and so on. In other words, for pressure P and volume V, at constant temperature T, $PV = k$ (a constant). This is the equation that became known as Boyle's Law.

Boyle was aware that a gas also expands when it is heated, but he had no reliable way of measuring temperature and so was unable to establish just what the relationship was between the volume of a gas and its temperature. It would be another hundred years before Jacques Charles (see page 33) succeeded in doing so.

UP, UP, AND AWAY

Towards the end of the 1600s, the French physicist Guillaume Amontons (1663–1705) developed the air thermometer. This measured temperature based

Boyle invented an improved air pump that greatly helped with his studies of gases.

on a proportional change in pressure, a relationship known as Amontons' law: $P/T = $ a constant. According to Amontons' law, increasing the temperature of a fixed volume of gas will increase its pressure. Amontons' law explains why you are recommended to adjust the pressure of your car tyres before setting out on a long trip. The friction of the tyre on the road surface raises the temperature of the air inside the tyre and, as a result, the air pressure inside the tyre increases.

On 15 October 1783, brothers Joseph and Etienne Montgolfier made a remarkable achievement. They launched a tethered balloon into the air, sent aloft by hot air. Suspended beneath the balloon was Jean-François Pilâtre de Rozier, a chemistry and physics teacher. He stayed aloft for almost four minutes, the first man to fly. About a month later, Pilâtre de Rozier, this time accompanied by the Marquis d'Arlandes, a French military officer, made the first free ascent. Their 25-minute flight took them some 9 kilometres out from their launch point in the centre of Paris.

While the Montgolfiers flew on hot air, mathematician, physicist and inventor Jacques Alexandre César Charles (1746–1823) found another way to go ballooning. Within days of the first free flight Charles launched the first manned hydrogen balloon with great ceremony from the Jardin de Tuileries in Paris. He was accompanied by Nicolas-Louis Robert on a flight that lasted just over two hours, covered 36 kilometres and reached an altitude of over 500 metres. Benjamin Franklin, who witnessed the beginning of the flight, wrote that: 'We observed it lift off in the most majestic manner... the intrepid voyagers lowered their hats to salute the spectators. We could not help feeling a certain mixture of awe and admiration.'

Charles and Robert took a barometer and a thermometer aloft with them which they used to measure the pressure and the temperature of the air at altitude. This meant that not only was this the first flight of a manned hydrogen balloon but also the first flight of a weather balloon.

Robert Boyle apparently first discovered hydrogen gas in the course of experiments carried out with iron and acids in 1671. Henry Cavendish (1731–1810) was the first to recognize it as an element, calling it 'inflammable air'. It was soon established that hydrogen was substantially less dense than air, making it an obvious if dangerous choice for a lighter-than-air balloon.

Inspired by his ballooning exploits, Charles carried out experiments around 1787 in the course of which he observed that the volume of a gas is directly proportional to its temperature – V/T = a constant. In principle, this was basically the same finding that Amontons had made some years earlier. This relationship between the temperature and volume of a gas, which became known as Charles' Law, explains how hot-air balloons work. Since at least the 3rd century BC, and Archimedes' 'Eureka' moment, it had been known that an object that weighs less than the fluid it displaces will float. Given that a gas expands when heated, a given mass of hot air will occupy a larger volume than the same mass of air when cold. So the hot air is by definition less dense than cold air. When a sufficient quantity of hot air is trapped in

The Hindenburg disaster in 1937 illustrated the danger of hydrogen balloons.

a balloon to make the total density less than that of the surrounding air the balloon begins to rise, floating on the cooler, denser air around it, like a cork bobbing on a pond.

Charles didn't publish his findings and it wasn't until 1801 that Joseph Louis Gay-Lussac (1778–1850) demonstrated the validity of Charles' Law for a number of different gases through careful experiment. When Gay-Lussac published his findings, he gave full credit to Charles' earlier findings and ensured that the law established was named in his honour. At the same time, Gay-Lussac also laid claim to a gas law of his own.

Gay-Lussac was a keen balloonist who ascended to record heights to test his theories on the composition of the atmosphere.

THE PRESSURE LAW

Gay-Lussac restated what is known as the pressure law, which states that if the mass and volume of a gas are kept constant the pressure of the gas will vary according to its temperature. Because Amontons had established the general principle of the pressure law a century earlier it is more properly known as Amontons' Law. When Gay-Lussac's law is mentioned today it more usually refers to what is known as the law of combining volumes, which states that, when gases react together to form other gases, and all volumes are measured at the same temperature and pressure, the ratio between the volumes of the reactant gases and the products can be expressed in simple whole numbers.

THE COMBINED GAS LAW

It was obvious that the gas laws were all linked. Boyle's Law linked pressure and volume, Charles' Law linked volume and temperature, and Amontons' Law linked temperature and pressure. The three can be written together as the combined gas law as PV/T = a constant.

A gas that obeys the combined gas law is referred to as an ideal gas. An ideal gas has a number of properties:

1. An ideal gas consists of a large number of identical molecules.
2. The individual molecules themselves take up no volume.
3. The molecules obey Newton's laws of motion, and they move randomly.
4. The molecules do not attract or repel each other; any collisions are completely elastic, and take a negligible amount of time.

Real gases behave in a way that is very close to the way the theoretical ideal gas behaves. A complete description of the behaviour of a real gas would come with the development of atomic theory by John Dalton and Amedeo Avogadro (see chapter 5).

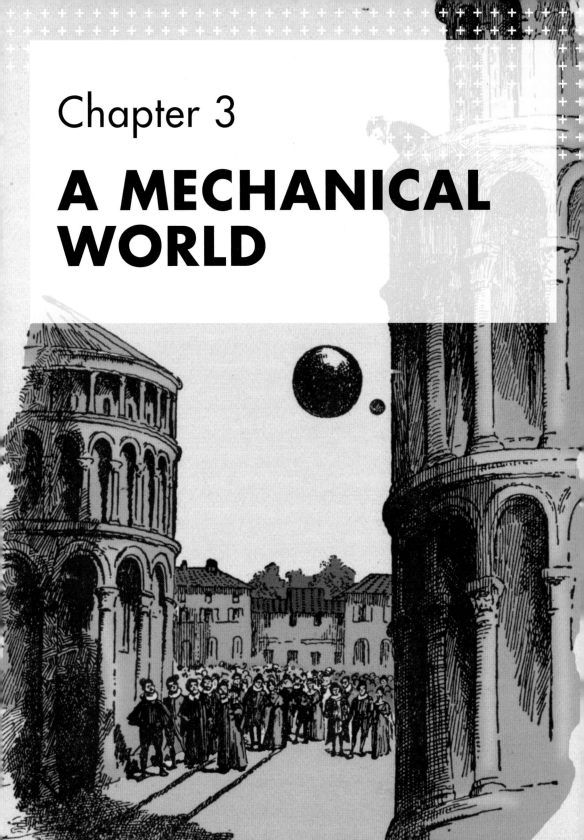

Chapter 3
A MECHANICAL WORLD

A Mechanical World

TIMELINE OF LAWS OF MOTION

Timeline	
4TH CENTURY BC	Aristotle proposes that objects will only move when they are propelled by a force, and will stop moving when that force is removed.
11TH CENTURY AD	The Persian scholar Avicenna argues that an object in motion will keep moving because of its 'impetus', imparted by the initial force that sets it in motion.
1543	Nicolaus Copernicus places the sun, not the earth, at the centre of the universe in his book *De revolutionibus orbium coelestium libri VI*.
1609–18	Johannes Kepler establishes his three laws of planetary motion.
1630s	Galileo discovers that falling objects accelerate uniformly, regardless of their mass and sets out his ideas on relative motion and on the path of projectiles.
1644	René Descartes introduces the idea of momentum, later expanded upon by Christiaan Huygens to take into account direction of movement as well as speed.
1687	Isaac Newton publishes the *Principia Mathematica*, in which he set out his three laws of motion and his law of universal gravitation establishing that objects attract one another with a force that varies as the product of their masses and inversely as the square of the distance between them.

Why do things move? On the face of it, it seems like a simple enough question. The Greeks, such as Aristotle believed, logically enough, that things will only move if propelled by a force and will come to a halt if that force is removed. Heavy objects fell because they were seeking out

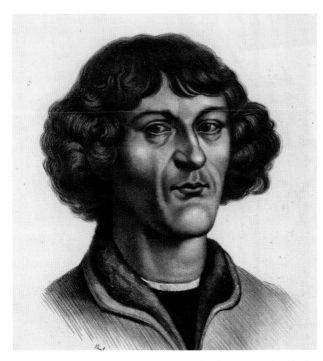
Copernicus is said to have received a copy of his book on his deathbed.

their natural place, which also explained why smoke would rise. All other kinds of motion required the application of a force – your plough wasn't going to go anywhere if you stopped pulling it. Yet there were obvious problems with this view. Everyone could see that a discus, for example, continued on through the air for some time after it left the thrower's hand. Aristotle tried to get around these difficulties by suggesting that the air itself supplied the propulsive force. Despite such shortcomings, Aristotle's view was largely unchallenged for 2,000 years.

In the Middle Ages, scholars such as the Persian Avicenna (980–1037) and Jean Buridan (1300–58) in France began to part company with Aristotle's teachings. They asserted that when a body is set in motion by a force, it keeps moving due to what was termed its 'impetus'. The impetus that is imparted to the object by the initial force, stays with it, and motion only ends when another opposing force counteracts the impetus. On the face of it this is similar to the modern idea of momentum, the natural tendency of an object in motion to keep on moving, but Avicenna and Buridan were viewing impetus as an internal force, actively pushing the object forward.

EVERYTHING UNDER THE SUN

In 1543, the publication of *De revolutionibus orbium coelestium libri VI* ('Six Books Concerning the Revolutions of the Heavenly Orbs') by the Polish astronomer Nicolaus Copernicus threw a cosmic spanner into our view of the world and set the scientific revolution in motion. Copernicus pointed out that calculating the positions of the planets in the night sky would be much simpler if the sun, rather than the earth, was assumed to be at the centre of the universe. One of the practical problems (along with the deep-seated philosophical ones) thrown up by the Copernican view was this: If the earth was not the still centre of creation, but was in

The heliocentric universe of Copernicus was a revolution in both science and society.

fact hurtling pell-mell through space, why weren't we aware of this motion? The very idea seemed to go against common sense. Word of the new Copernican model of the universe spread slowly. *On the Revolutions of the Celestial Spheres* was placed on the Catholic Church's list of banned books and would stay there until 1835. Before Copernicus, astronomers had tried to account for the observed motions of the stars and planets by imagining that they fixed to crystal spheres centred on the earth. Copernicus still believed that the universe was formed of these nested perfect spheres, just that they were no longer centred on the earth.

At the beginning of the 17th century, the German astronomer Johannes Kepler made a series of painstaking observations of planetary motion that led him to a sensational conclusion. The paths that the planets followed on their journeys across the sky were not

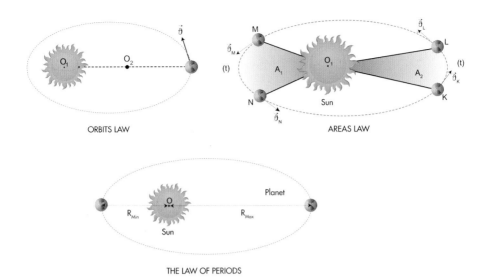

ORBITS LAW

AREAS LAW

THE LAW OF PERIODS

Kepler's Three Laws of Planetary Motion.

perfect circles. Rather they were moving in flattened circles, or ellipses. After Galileo's discovery of the moons of Jupiter, Kepler found that they, too, moved in elliptical paths around the giant planet.

Beginning with the first law, the law of orbits, in 1609, Kepler set out his three laws of planetary motion (the second and third laws, the law of equal areas and the law of periods would follow in 1618). These laws described *how* the planets moved, but not *why* they moved. Kepler tried to work out what force might be responsible for the planets moving as they did. He thought that magnetism might be involved and that the sun must have something to do with it but couldn't come up with a satisfactory explanation. That would come 70 years later, from Isaac Newton.

GALILEO

Italian mathematician and scientist Galileo Galilei was born in Pisa, Italy. As a young man, he studied medicine at the University of Pisa but soon became interested in mathematics and physics. Specifically, he was curious about the way objects behaved when they moved. Opinion and argument, Galileo believed, were not enough. Through his work, Galileo helped develop the modern scientific method, establishing that ideas can only be proved or disproved by careful experiment. In his *Dialogo sopra i due massimi sistemi del mondo* (*Dialogue Concerning the Two Chief World Systems*), published in 1638, Galileo's intent was to defend the idea that the earth does not sit motionless at the centre of the universe.

Galileo was indisputably one of the world's greatest scientists.

In the 1630s, Galileo studied the motion of balls rolling on inclined planes. He noticed that, if he rolled a ball down one plane and up another, it would reach the same height on the second plane as it had started out on from the first no matter how steeply inclined his planes were. He reasoned from this that if the second plane were horizontal the ball would keep rolling forever unless something stopped it. This was contradictory to the Aristotelian notion that objects needed a force to stay in motion. Galileo realized that there was no real difference between an object moving at a steady speed

Galileo determined that experiments carried out onboard a smoothly moving ship would give the same results as those carried out on land.

and direction and an object that wasn't moving at all. Neither one was being subjected to outside forces. In setting out these ideas Galileo was introducing the principle of inertia, which would be developed by Isaac Newton some 50 years later.

Galileo's thinking also anticipated that of another great scientist of later years, Albert Einstein. Galileo explored the idea that all motion is relative and that it only makes sense to speak of something moving in relation to something else. Galileo tackled the argument by imagining the situation of a passenger aboard a ship sailing on a perfectly smooth lake at a constant speed. He asked, is there any way in which the passenger can determine that the ship is moving without going on deck? If the ship continues to move at constant speed and direction the passenger will not feel its motion, just as a modern-day traveller doesn't notice the motion of an aircraft cruising through the sky.

Galileo concluded that any mechanical experiment performed inside the ship, always provided that it was moving at constant speed in a constant direction, would give exactly the same results as a similar experiment carried out on shore, and therefore give positive proof that the ship was in motion. From these observations Galileo put forward his own relativity hypothesis:

Any two observers moving at constant speed and direction with respect to one another will obtain the same results for all mechanical experiments.

This, then, was the reason why the earth's motion through space could not be discerned. The earth and everything upon it, including all the experimenting scientists and vexatious clergymen, were all in the same state of motion relative to each other.

UNIFORM ACCELERATION

Galileo wasn't done there. Aristotelian thought also asserted that a heavier object would fall faster than a lighter one. Again, it seemed to make sense, but Galileo was more interested in experimental proof. Rolling a ball down an inclined plane, rather than simply dropping it, allowed Galileo to slow the rate of acceleration sufficiently for him to be able to time it (not having a clock for this purpose Galileo used his own pulse, or the swinging of a pendulum, as a timekeeper). Timing the rate of descent of differently weighted balls and taking into account the slowing effect of friction, Galileo concluded that freely falling objects accelerated uniformly at a rate that was independent of their mass. More exactly, he demonstrated that the distance travelled by a constantly accelerating object is proportional to the square of the time in which it has been falling. That a feather might fall more slowly than an egg was simply a consequence of the greater effect of air resistance on the former.

In *The Dialogues* Galileo wrote, 'It has been observed that missiles and projectiles describe a curved path of some sort; however, no one has pointed out the fact that this path is a parabola.' He understood that the motion of any projectile was the resultant of two forces acting upon it, the initial force that set it in motion and the force that pulled it towards the ground. The combination of the two forces meant that the projectile must follow a parabolic path.

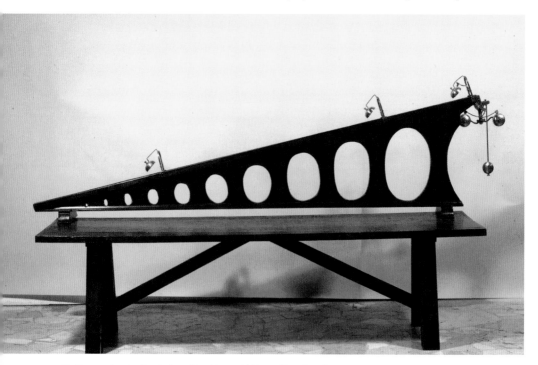

Galileo used an inclined plane like this to aid his studies of motion.

Galileo's discoveries set the groundwork for what is now referred to as classical mechanics, they embraced not only falling bodies and projectiles but also took in an analysis of the motion of the pendulum. He laid a path for Isaac Newton to follow on the way to his laws of motion, which would incorporate Galileo's ideas, and the theory of gravity.

THE CONSERVATION OF MOMENTUM

The idea of momentum, or the amount of motion of an object, was introduced by French philosopher René Descartes in his *Principia Philosophiae* (*Principles of Philosophy*) in 1644. The momentum of a moving object was simply the product of its mass and its speed.

That the total momentum is conserved in any system of interacting objects is one of physics' great laws. At first, Descartes couldn't make it work. Imagine two objects of the same size, travelling at the same speed, but in opposite directions, colliding and coming to a halt. It seems obvious that each object had momentum when it was in motion but when they came to a stop their momentum apparently became zero. This rather foxed Descartes. It was Dutch physicist Christiaan Huygens (1629–95) who suggested that it was necessary to take into account not only the object's speed, but also its direction. In other words, its velocity. If an object moving in one direction was considered to have positive momentum then an object moving in the opposite direction would have negative momentum. A system consisting of two objects of equal mass moving together from opposite directions at the same speed would have total momentum of zero as the momentum of one cancelled out the momentum of the other. In other words, when they collide and come to a halt, the total momentum after the collision is the same as the momentum afterwards – zero.

Momentum is traditionally denoted by the letter P, so the definition of momentum can be written as $P = mv$ (where the object has mass m and is moving at velocity v). P and v are both vector quantities, that is they have magnitude and direction.

LET NEWTON BE!

In a room in the house where Isaac Newton was born in Woolsthorpe, Lincolnshire, a marble tablet is inscribed with these words by the poet Alexander Pope:

Nature and Nature's laws lay hid in night
God said: Let Newton Be! – and all was light

Until Albert Einstein's relativity revolution at the start of the 20th century, our understanding of the laws that govern the movement of objects through space was founded on the work of Isaac Newton. The work done previously by Galileo came under the branch of classical

René Descartes introduced the idea of momentum.

mechanics known as kinematics, or the description of motion. But Galileo and others, such as Kepler, had not succeeded in coming up with explanations of the *causes* of motion. Real success in this field of mechanics, called dynamics, would come with Newton. He brought together Galileo's terrestrial mechanics of falling bodies and linked it with the celestial mechanics of Kepler and produced a unified system of movement.

In 1687, Newton produced (after a long delay – it is believed that he may have formulated its central ideas around 1666) what is believed by many to be a contender for the greatest work of science ever written. The *Philosophiae Naturalis Principia Mathematica* (*Mathematical Principles of Natural Philosophy*), generally referred to simply as the *Principia*, set out Newton's vision of a universe in motion. It was here that he formulated his three laws of motion, which were concerned with the motion of objects and the forces acting upon them:

1: An object will remain at rest or continue to move in the same direction and at the same speed unless acted on by a force.

2: A force acting on an object will cause it to move in the direction of that force. The magnitude of the change in the speed or direction of the object is dependent on the size of the force and the mass of the object.

3: For every action there is an equal and opposite reaction. If one object exerts a force on another, an equal and opposite force is exerted by the second object on the first.

THE UNIVERSAL FORCE

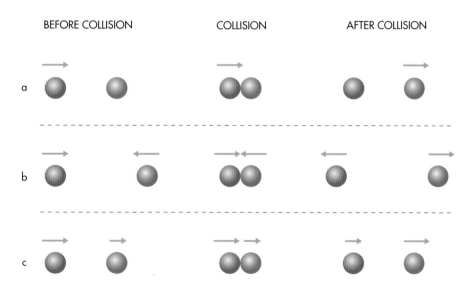

Conservation of momentum

Isaac Newton's laws of motion determined why objects move as they do.

The story of Newton in the apple orchard is an often repeated one, but it really was a singular mind that asked the question: 'If an apple falls to the earth, why doesn't the moon?' It took real genius to conclude that the moon actually *is* falling.

Newton knew that any explanation he came up with to explain the motion of both the apple and the moon would have to explain Kepler's findings too. Newton determined that between any two objects there is always a gravitational force that attracts them to each other. The strength of the force depends on the masses of the objects and on the distance between them. Gravity obeys an inverse square law, which means that the magnitude of the force decreases by the square of the distance. Therefore, if you double the distance between two objects the force that draws them together reduces to just a quarter of what it was. At five times the distance the force is reduced to a 25th of what it was. The law of Universal Gravitation, therefore, states that objects attract one another with a force that is proportional to the product of their masses and inversely proportional to the square of the distance between them.

With three simple laws of motion and one law of gravity, Newton, it seemed, could explain the movement of everything in the universe. Newton's laws provided an explanation for Kepler's laws of planetary motion and for the fall of the apple. Newton derived his laws from three fundamental quantities that underpin all of science – time, mass and distance. Knowing the time an object takes to travel a set distance enables you to calculate its velocity (speed and direction). Mass tells you how much matter that object contains and therefore what amount of force you'll need to move it. Multiply the mass by the velocity and you'll get the object's momentum, which will tell you how hard it's going to be to stop it.

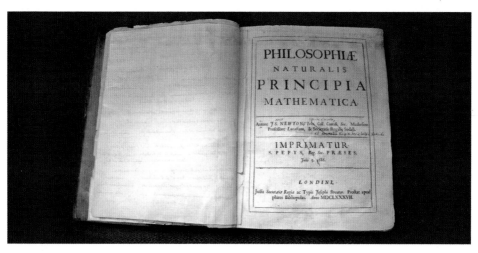

Newton's Principia, *a contender for science's greatest work.*

CANNONBALLS AND SATELLITES

As established by Galileo, two forces govern the path of a projectile – gravity and the initial force imparted to it that sends it on its way. The result of those two forces acting on it is that the projectile follows a curved path back to earth. Newton imagined mounting a powerful cannon atop a mountain. Ignoring inconveniencies like air resistance, he knew that the distance (d) the cannonball travelled horizontally would be determined by its speed (v) multiplied by the time (t) it spent in flight – d = v x t. He also knew from Galileo's work with falling objects that the time spent in flight would be determined by how long it took gravity to pull the cannonball to the ground.

Newton looked at the two formulas for the distance a cannonball would travel horizontally and vertically. He noticed that the distance the cannonball would fall in a given time was constant, since the acceleration due to gravity is constant, but, the distance the cannonball travelled horizontally was dependent on its speed. Changing the speed at which the cannonball was fired changed its trajectory.

Newton realized that if he chose just the right velocity, the curved trajectory followed by the cannonball would exactly match the curvature of the earth's surface. It would always stay the same height above the ground. The inertia of the cannonball (which makes it want to continue travelling in a straight line) is balanced against the acceleration due to the earth's gravity (which pulls the cannonball towards the centre of the earth). The cannonball will now travel right around the earth, always accelerating towards the planet but never reaching the ground. It is now a satellite in orbit.

This is exactly the principle that puts actual satellites into orbit, except with powerful rockets rather than a cannon providing the forward motion. The moon is like a gigantic cannonball, perpetually falling around the earth in its orbit.

Newton's laws were unchallenged for more than 200 years. For everyday purposes, they are still an excellent way of calculating the movement of an object and how it is affected by the force of gravity. But still Newton had failed to explain what actually caused gravity. An answer to that would have to wait for Albert Einstein.

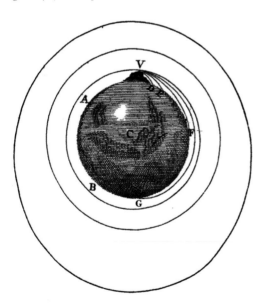

Newton's cannonball thought experiment predicted the orbits of today's satellites.

Chapter 4
SEEING THE LIGHT

Seeing the light

TIMELINE OF THEORIES OF LIGHT

Timeline	
6TH CENTURY BC	Pythagoras believes that the eyes produce invisible rays allowing us to see objects.
4TH CENTURY BC	Democritus suggests that objects emit images of themselves and that colour results from the roughness of the atoms that make them up.
3RD CENTURY BC	Euclid writes *The Optics*, in which he argues that light rays emanate from the eyes and travel in straight lines.
1ST CENTURY BC	Lucretius argues that light is made up of minute particles that travel too fast to be detected individually.
11TH CENTURY	Al-Haytham establishes that we see things because light reflects from them and that light has a very high, but finite, speed. He sets out his ideas in *The Book of Optics*.
17TH CENTURY	Johannes Kepler develops the theory of the camera obscura and explains how the human eye works for the first time.
1690	Christiaan Huygens publishes *Treatise on Light* proposing the wavelike nature of light and that it travels through an invisible 'aether'.
1704	Isaac Newton publishes *Optiks*, which includes an account of his experiments with prisms. He firmly believes light is a stream of particles.
1801–3	Thomas Young demonstrates that light forms interference patterns, strongly suggesting its wavelike nature.
1819	Augustin-Jean Fresnel presents his theory of diffraction, proving light is a wave, to the French Academy of Sciences.

For centuries, people have tried to explain the various phenomena associated with light. The 6th century BC Greek philosopher Pythagoras (c.569–c.499 BC) thought sight was like a very delicate sense of touch, the eyes producing invisible rays with which we sense objects. Another Greek thinker, Democritus (c.460–370 BC), believed that objects continually emitted images of themselves, which we sensed. He was one of the first to attempt to explain perception and colour, suggesting that sensation was caused by the size and shape of atoms, and that colour was caused by properties such as the roughness of the atoms. Plato (428–348 BC) put forward the idea that the inner light from the eyes had to mix with the light from the sun before we could see anything. Aristotle rejected the idea that the eyes were the source of light, believing that objects emitted light and that this was detected by the eye. He also believed that the watery surface of the eye created a kind of screen on to which light was projected.

In a textbook called *The Optics,* Euclid (c.325 –c.265 BC) suggested that visual rays were straight lines and defined the apparent size of an object in terms of the angle formed by the rays drawn from the top and the bottom of the object to the observer's eye. He also maintained the idea that light rays emanate from the eyes, though it did occur to him to wonder why, when you looked up into the night sky, the stars were visible instantaneously. Assuming that the light beams travelled infinitely fast offered an answer to this problem. Possibly the fact that his geometrical analysis of the way light travels was so successful and useful helped to ensure that the notion of light from the eyes held sway for centuries.

WAVES OR PARTICLES?

Notwithstanding what they believed its source to be, the Greeks also had opposing views as to the nature of light. Aristotle suggested that light was a disturbance in the ether, an invisible, undetectable substance that filled space. He perceived light as a wave that travels through the ether like an ocean wave travels across the water. Another view held that light was a stream of tiny particles that were too small and fast-moving to be perceived individually. The Roman philosopher Lucretius wrote in *On the Nature of the Universe* (55 BC): 'The light and heat of the sun... are composed of minute atoms which, when they are shoved off, lose no time in shooting right across the interspace of air in the direction imparted by

The title page from a medieval edition of Lucretius' De Rerum Natura.

the shove.' As the physics of a later era would show, this was a remarkably prescient view, but Lucretius' ideas were not generally accepted and the particle theory was opposed by those who continued to champion Aristotle. For the next two thousand years or so, it was generally accepted that light travelled in waves.

THE MEDIEVAL VIEWPOINT

The Arabian physicist Ibn al-Haytham (965–1038), also known as Alhazen, finally put to rest the idea that beams of light emanated from the eyes. He established once and for all that we see things either because they reflect light from a source of illumination, or they are themselves a source of illumination, whether it be a candle or the sun. Al-Haytham, who has sometimes been described as 'the father of optics', argued that we see objects because the sun's rays of light, which, like Lucretius had suggested earlier, he believed to be streams of tiny particles travelling in straight lines, were reflected from objects into our eyes. He understood that light must travel at a large but finite velocity, and that the phenomenon of refraction was caused by the velocity being different in different substances. He also understood how

it was the refraction of light by a lens that allowed images to be focused and magnified. He set out his ideas in *The Book of Optics*, written while he was under house arrest in Cairo from 1010–1021, having fallen foul of the caliph after foolishly promising that he could regulate the Nile floods.

Parts of *The Book of Optics*, translated into Latin, reached Europe around 1200. The English scholar Robert Grosseteste (*c*.1168–1253) was one scholar who read al-Haytham's work and carried out some experiments of his own. He believed that the entire universe had been formed from light. Light was the first of all things to be created, expanding out from a single point into a

An illustration from al-Haytham's Book of Optics *shows ideas on phenomena such as perspective, reflection and refraction.*

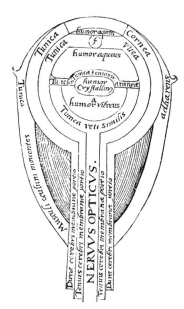

Ibn al-Haytham's diagram of the anatomy of the eye.

sphere that contained all other things within itself. This was a startlingly sophisticated notion that has obvious parallels with our current thinking of how the universe formed.

Roger Bacon was an English monk and a pupil of Grosseteste's who shared his teacher's enthusiasm for studying light. By some accounts, Bacon was the first modern scientist (although al-Haytham might also lay claim to that title), placing great emphasis on the importance of carrying out experiments. Some of these experiments involved bending and focusing light by passing it through a lens. Bacon was one of the first to suggest spectacles for people with poor eyesight.

Scholars discussing optics in Europe at the time were sometimes referred to as 'perspectivists' after Roger Bacon's influential book *Perspectiva* from around 1270. This medieval idea differed from the idea of perspective in painting, and referred to the science of seeing itself. The perspectivists thought that optics played a big role in the way we gathered knowledge of the world. It begins with the emission of light and colours from the objects through air to the eyes. This was a view based on al-Haytham's book. Bacon believed that the essence of a visible object entered the eye and reproduced itself there.

The ideas of Ibn al-Haytham were not improved on for over four hundred years. The first to take a real step forward was Johannes Kepler. He provided the first correct mathematical theory of the camera obscura, the projecting of an image through a pinhole on to a screen in a darkened room. He gave the first explanation of how the human eye works, with an upside-down image formed on the retina, an idea that was widely rejected on the not unreasonable grounds that we do not see the world upside down. Kepler also correctly explained what caused short sight and long sight. In addition, he made the calculation that the intensity of light varied inversely with the square of

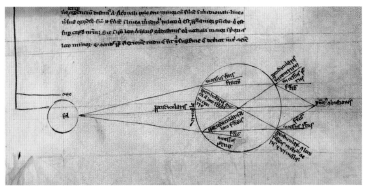

Bacon's illustration of light refracting through a glass sphere filled with water.

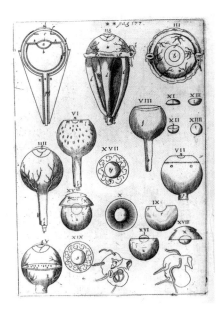

Astronomiae Pars Optica *(1604) by German astronomer Johannes Kepler was one of the first detailed anatomical studies of the eye.*

the distance of the observer from the source. He also argued that the velocity of light was infinite, in which assumption he was mistaken.

LIGHT AND COLOUR

For centuries, light and colour had been held to be two different phenomena. Colour was a property inherent in an object that was carried from it to the observer by the light. Light was a carrier of colour rather than the source of colour itself. René Descartes suggested around 1637 that colour might be caused by the rotation of the particles that formed a beam of light and that therefore colour was a property of the light itself.

Isaac Newton carried out a series of ingenious experiments that did much to reveal the nature of light, which he published in *Opticks* in 1704. One of the best known of these was the refracting of a beam of white light through a prism to reveal the colours of the spectrum. His revelation that white light could be split into a rainbow of colour and that each colour is refracted by a different degree was conclusive proof that colour was indeed a property of light.

Isaac Newton's experiments with prisms proved conclusively that colour is a quality of light.

Newton was a firm believer that light was a stream of minute particles and that the refraction of the light through the prism represented particles of different sizes being deflected to a greater or lesser degree as they passed through. He also studied the partial reflection of light from transparent materials, attempting to account for this by suggesting that the light particles might sometimes travel in 'fits and starts', making them sometimes liable to be reflected rather than transmitted.

Another important observation involved what came to be named Newton's rings. Newton noticed that when he placed a convex lens on a flat glass surface, so there was a thin layer of air between the two, he could see a series of concentric light- and dark-coloured bands. Newton surmised that this was due to the particles of light being set vibrating back and forth between the glass surfaces. The phenomenon is actually caused by light waves reflected from the top and bottom surfaces of the air film interfering with each other but Newton was reluctant to accept that light had wavelike properties.

Such was the strength of Newton's influence on the scientific world that the corpuscular, or particle, theory of light became the generally accepted one, but it was by no means everyone who agreed with him.

MAKING WAVES

Francesco Maria Grimaldi (1618–63), an Italian physicist, had examined the transmission of light through a small hole and observed that a small amount of light could be seen in regions that would have been expected to be in shadow had it followed the straight line course that a stream of particles would have been expected to take. He called this phenomenon diffraction, from the Latin verb *diffringere* 'to break into pieces'. Grimaldi also observed that the light pattern consisted of complicated coloured bands. On the basis of the evidence before him he speculated that light might travel in a wavelike manner.

Scottish physicist and optician James Gregory (1638–75) gave an excellent description of a 'diffraction grating' in a letter of May 1673 that he intended should be communicated to Newton:

Francesco Grimaldi's observations of diffraction led him to believe light was wavelike.

If ye think fit, ye may signify to Mr. Newton a small experiment, which (if he know it not already) may be worthy of his consideration. Let in the sun's light by a small hole to a darkened house, and at the hole place a feather, (the more delicate and white the better for this purpose,) and it shall direct to a white wall or paper opposite to it a number of small circles and ovals, (if I mistake them not) whereof one is somewhat white, (to wit, the middle, which is opposite to the sun,) and all the rest severely coloured. I would gladly hear his thoughts of it.

In *Opticks*, Newton wrote about what he called inflexions (his name for diffraction). He described the use of a feather as diffraction grating, as Gregory had done, but made no mention of Gregory himself. He wrote that: 'When I made the foregoing Observations, I design'd to repeat most of them with more care and exactness, and to make some new ones for determining the manner how Rays of Light are bent in their passage by Bodies, for making the Fringes of Colours with the dark lines between them. But I was then interrupted, and cannot now think of taking these things into farther consideration.' This might be regarded as being a not particularly useful response. He set out a series of questions for others to pursue but never returned to the study of optics himself.

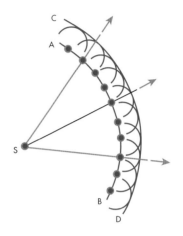

Huygens' Principle: *Each wavefront is the envelope of the wavelets. Each point on a wavefront acts as an independent source to generate wavelets for the next wavefront. AB and CD are two wavefronts.*

Christiaan Huygens believed in the wavelike nature of light. He worked out the mathematical details of his theory in 1678 and published it in his *Treatise on Light* in 1690. Of course a wave needs a medium to travel through, as, for example, sound waves travel through air, so Huygens proposed that light travelled through an invisible, but all-pervading ether. He theorized that every point on a wavefront could be considered as a source of secondary spherical wavelets which spread out in the direction of travel of the light wave at the speed of light. It was an elegant theory and explained most of the observed phenomena of light such as reflection, refraction and diffraction; unfortunately it was more or less ignored. Others, such as Swiss mathematician Leonhard Euler (1707–83), argued strongly for a wave theory of light but it wasn't until the early years of the 19th century that Thomas Young produced major experimental evidence that light did indeed have wavelike properties.

PHENOMENON YOUNG

Thomas Young (1773–1829) was a child prodigy and a man of such formidable intellect that his fellow students at Cambridge University dubbed him 'Phenomenon'. He was said to have read the Bible from beginning to end twice by the age of six. He was an accomplished linguist

Thomas Young was a formidably talented scientist who challenged Newton's ideas about light.

and played a role in the deciphering of the Rosetta Stone, which finally allowed archaeologists to decode Egyptian hieroglyphics.

Between 1801 and 1803 he delivered a series of lectures to the Royal Society in London. In his talk of 1801, he put forward his theory of three-colour vision to explain how the eye detected colours. It was an idea that wouldn't be confirmed until the 1950s. In the course of another talk in 1803 he described an experiment that was beautiful in its elegance and simplicity.

In 1801, Young had described an effect he called interference. If two waves meet they don't bounce off each other like colliding snooker balls; instead, they appear to pass straight through each other. Watch raindrops falling on to a pond and see how the ripples spread and meet and keep on going as they cross each other. Where the waves cross they combine with each other. If the peak of one wave meets the peak of another they are added together to make a higher peak; two troughs make a deeper trough and a trough and a peak cancel each other out. The result was an interference pattern that showed where the waves were adding and cancelling. Young hypothesized that if light was wavelike then it should behave in a manner that was similar to the ripples on the pond.

He began by making a small hole in a window blind, which gave him a point source of illumination. Next, he took a piece of board and made two pinholes in it, placed close together. He positioned his board so that

Illustrations from Young's lecture notes showing things such as diffraction and the inverted image in the retina.

the light coming through the hole in the window blind would pass through the pinholes and on to a screen. If Newton was right, and the light was a stream of particles, then there would be two points of light on the screen where the particles travelled through the pinholes.

Rather than two discrete points of light, Young saw a series of curved, coloured bands separated by dark lines, exactly as would be expected if light were a wave. Young called these bands interference fringes. Young's experiment appeared to be convincing proof of the wave nature of light. He theorized that if the wavelength of light was sufficiently short it would explain why it appeared to travel in straight lines, as if it were a stream of particles. Unfortunately, it was not really the done thing to contradict the great Newton and Young's findings were not well received by his peers.

FRESNEL'S MODEL

It would be another 15 years before French physicist and engineer Augustin-Jean Fresnel (1788–1827) appeared to have demonstrated conclusively that light really was a wave. Fresnel carried out experiments on diffraction that were similar to those of Grimaldi and Newton. Using a small lens to collect sunlight, and making observations through an eyepiece, he was able to study the phenomenon of diffraction in great detail. He believed the light to propagate in the way described by Huygens to produce both diffraction and interference effects. The French Academy of Sciences had announced a competition to

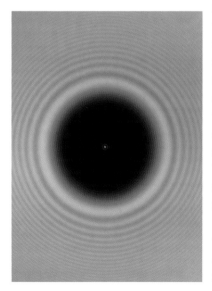

settle the disagreement between the particle and wave theories of light and Fresnel submitted a mathematical description of his observations. One member of the prize committee, the mathematician Simeon-Denis Poisson, noted that Fresnel's theory, if correct, indicated that a bright spot would be seen in the centre of the shadow cast by an opaque circular disc. Poisson believed that there would be no such spot and wave theory would be shown to be wrong. When the experiment was carried out, there was the spot. Fresnel won the prize and time seemed to be up for the corpuscular theory of light.

But still the question remained. What actually *was* light? The beginnings of an answer would come from the study of a seemingly unrelated force – electricity. And as science began to explore the quantum realm in the 20th century, Thomas Young's two-slit experiment would resurface as one of the most profound in all of physics.

The tiny bright spot at the centre of the image was experimental proof of Fresnel's idea of the wave nature of light.

Chapter 5
STATES OF MATTER

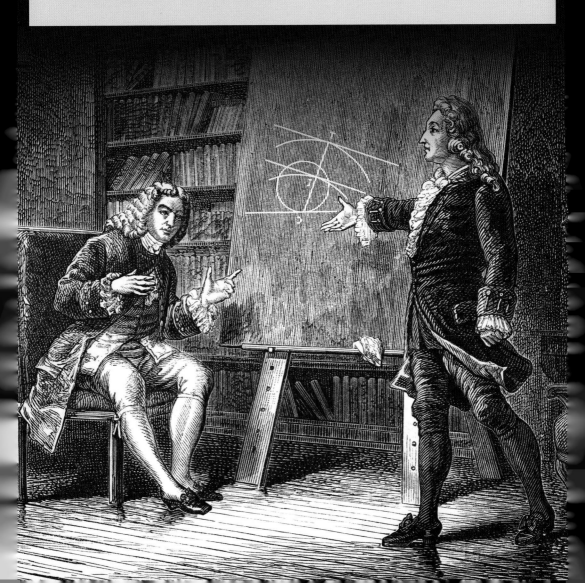

States of Matter

TIMELINE OF THEORIES OF MATTER

Timeline	
5TH CENTURY BC	Empedocles argues that all matter is made up of four primal elements – earth, air, fire and water.
5TH CENTURY BC	Democritus suggest that matter is formed from infinitesimally small particles called *atomos*.
1661	Robert Boyle theorizes that matter is made of corpuscles that can be rearranged to form different elements.
1738	Daniel Bernoulli proposes a kinetic theory of gases, suggesting that the pressure of a gas is proportional to the kinetic energy of its particles.
1773	Antoine Lavoisier states that mass is neither created nor destroyed in a chemical reaction.
1803	The first scientific theory of the atom is put forward by John Dalton.
1811	Amedeo Avogadro hypothesizes that equal volumes of gases at the same temperature and pressure contain equal numbers of molecules.

People became adept at making use of matter, carving wood, creating ceramics and working metals for example, long before they understood what it actually was. The Greek philosopher Thales is generally credited as being the first person who tried to answer the question: what is everything made of? His answer was, perhaps, not obvious. The universe, he decided, was made of water. Others disagreed, asking not unreasonably where the water was in a hot, dry rock. Empedocles (492–432 BC) declared that all matter was composed of four primal elements – air, earth, fire and water – the ratio of those elements determining the nature of the matter. Unlikely as it might seem, Empedocles' theory did point the way to a valid scientific idea, that materials were rarely 'pure' but were made up of combinations of different things.

Empedocles.

Some time later, Democritus (460–370 BC), another Greek thinker, proposed a new theory of matter. Democritus' idea drew on the work of Anaxagoras, who had believed that matter was infinitely divisible ('Of what is small, there is no smallest part, but always a smaller,' he wrote in *About Nature*) and Leucippus, who suggested that matter consists of an infinite number of particles, which were indivisible but so small as to be invisible. Democritus knew that if you took a stone and cut it in half, each half had the same properties as the original stone – at no point would you even get a glimpse of the air, earth, fire and water constituents Empedocles said it was composed of. Democritus reasoned that if you kept on dividing the stone, eventually a point would be reached at which the stone was in pieces so tiny it would be physically impossible to divide it further. Democritus called these infinitesimally small pieces of matter *atomos*, meaning 'indivisible'. He suggested that *atomos* were eternal and could not be destroyed. He further suggested that each material was made up of its own specific *atomos* – the *atomos* of a stone were unique to it and distinct from the *atomos* of a feather, say.

This was a remarkably insightful view, arrived at purely theoretically, but others, particularly the influential Aristotle, did not agree with it and it was largely forgotten for two millennia. Aristotle believed that substances could be understood in terms of matter and form. Matter without form cannot exist. Matter, according to Aristotle, is the stuff out of which things are made while form is what gives a thing its shape and structure and determines its characteristics and functions. To Empedocles' four elements, Aristotle added a fifth – quintessence, or aether, a

Empedocles' suggestion that all matter is formed from air, earth fire and water was surprisingly long lasting.

Democritus, who first suggested that matter is formed from indivisible 'atoms'.

divine substance that formed the stars and planets. Each of the elements had its natural place to which it would always try to return, which, according to Aristotle, explained why rain fell and flames rose up, for example. Unlikely as it might seem to us, it was a viewpoint that would dominate scientific thinking into the Middle Ages and beyond.

ALCHEMY AND ATOMISM

From Roman times to the age of the Enlightenment in the 17th and 18th centuries, alchemy, today sometimes considered to be no more than pseudoscience on a par with astrology and the like, was an important and respected branch of enquiry into the way the world worked. It may, in fact, be better thought of as a 'proto-science'. Alchemists contributed to the development of a huge range of skills and knowledge including basic metallurgy, metalworking, and the production of inks and dyes. Alchemy took in physics and medicine, as well as mysticism and spirituality, and it also developed an understanding of chemical

Alchemists developed a wide range of skills and expertise in their pursuit of knowledge.

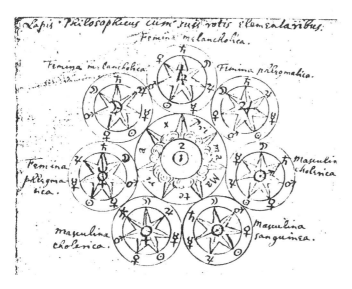

Isaac Newton was keenly interested in alchemy and wrote a great deal on the subject.

processes that gave rise to the modern science of chemistry. The goals of alchemy were to find the 'elixir of life' that would grant immortality; to find the 'philosopher's stone', which would turn 'base' (nonprecious) metals into gold, which was believed to be the highest and purest form of matter; and to uncover the role of humans in the cosmos and advance the human spirit.

Alchemy shaped the thinking of some of the prominent physicists of the 17th century, such as Robert Boyle and Isaac Newton. Newton's notebooks, for example, reveal his interest in the transmutation of metals and show that he wrote far more on the subject of alchemy than he ever did on physics. Boyle practised alchemy until the end of his life; he claimed to have witnessed a demonstration of the philosopher's stone and tested the gold it produced. He even successfully petitioned Parliament in 1689 to repeal a law forbidding gold-making, as he thought this was standing in the way of research into the powers of the stone.

Boyle was an advocate of a form of atomism. He defined elements in his work *The Sceptical Chymist*, published in 1661, as 'certain primitive and simple, or perfectly unmingled bodies... not being made of any other bodies, or of one another'. As a chemist, Boyle was interested in finding out what things were made of. He set aside the longstanding Aristotelian classical elements of air, fire, earth and water in favour of a theory of corpuscles. The corpuscular theory differed from the atomic theory of Democritus on the belief that corpuscles could, in theory, be divided, whereas atoms could not. Boyle believed that the transmutation of one element into another could take place through the rearrangement of the corpuscles that made up each element.

KINETIC THEORY

One of the findings that helped revive the atomic theory of matter was Torricelli's invention of the barometer in 1643 (see page 30), with which Boyle was familiar of course. In doing

Daniel Bernoulli, who outlined a kinetic theory of gases.

so, Torricelli had demonstrated that air had weight (it could push up a column of mercury) and must therefore be composed of something substantial. With his gas law (see page 31), Boyle had shown that air resisted compression and would expand to fill the available space. He had suggested two explanations for the way gases behaved: that air was composed of particles that were like coiled up springs and repelled each other; and that air was composed of particles in constant motion that were forever colliding with and bouncing off each other.

Swiss mathematician Daniel Bernoulli (1700–82) put forward a kinetic theory of gases in a book on hydrodynamics published in 1738. He derived Boyle's law for gas pressure by calculating the force exerted on a movable piston by the impacts of a number of particles moving with speed (v), in a closed space. The smaller the enclosed space the greater will be the pressure because the particles will strike the piston more frequently. Bernoulli also showed that the pressure will be proportional to the kinetic energy of the particles since the frequency of impacts is proportional to the speed of the particles and the force of each impact is proportional to the particle's momentum. This explained why increasing the temperature also increased the pressure. Bernoulli's kinetic theory introduced the idea that heat or temperature could be identified with the kinetic energy of particles. It had a rival in caloric theory, championed by the likes of Antoine Lavoisier (1743–94) and John Dalton (1766–1844), which suggested that gas molecules were propelled into the walls of a container by a 'heat substance', called caloric. It would be another hundred years or more before Bernoulli would win the day.

DALTON'S ATOMIC THEORY

Eighteenth-century chemists such as Joseph Priestley (1733–1804) and Antoine Lavoisier had demonstrated through their experiments that some substances could combine to form new materials, that some substances could be broken down into other materials, and that a few substances appeared to be 'pure' and couldn't be broken down any further. By careful measurement, Lavoisier had shown that when substances were burned the new materials formed weighed more than the original substances, and that this extra mass came from the air. From this, Lavoisier formulated the Law of Mass Conservation, which states that mass is neither created nor destroyed in the course of a chemical reaction.

In 1808, English scientist John Dalton put forward an overarching theory that drew previous findings together in a coherent whole. It would be the first truly scientific theory of the atom, arrived at through experimentation and analysis of the results.

Dalton had been led towards his theory through his research into gases. At the beginning of the 19th century Dalton had become secretary of the Manchester Literary and Philosophical Society, where he submitted a number of essays setting out his findings on such things as steam pressure at different temperatures and the thermal expansion of gases.

John Dalton's atomic theory put ideas about the atom on a firm scientific basis.

Based on his observations, Dalton concluded that all fluids (gases and liquids) under the same pressure expand equally as the temperature is increased. This became the basis for Dalton's Law, or Dalton's law of partial pressures, which stated that in a mixture of non-reacting gases, the total pressure exerted is equal to the sum of the partial pressures of the individual gases.

In the course of these investigations Dalton developed the idea that the sizes of the particles making up different gases must be different. He based his ideas on Lavoisier's law of mass conservation, according to which the mass of the reactants in a chemical reaction is the same as the mass of the products, and on French chemist Joseph Louis Proust's law of definite proportions from 1799, which stated that if a compound is broken down into its constituent elements, then the masses of the constituents will always have the same proportions, regardless of the quantity of the original substance. Experiments performed by Dalton and others such as Gay-Lussac (see page 68) indicated that the ratios between reactants and products in a chemical reaction were always simple whole numbers. To this, Dalton added his law of multiple proportions, stating that if two elements can be combined to form a number of possible compounds, then the ratios of the masses of the second element, which combine with a fixed mass of the first element, will be ratios of small whole numbers. Two elements might form a compound by combining at a ratio of 2 units to 1 unit, or 3 units to 2 units, but never 2.1 to 0.9, or 3.1 to 2.1.

In 1803 he set out his atomic theory, based on these ideas. The basic assumptions of the theory were that:

1. All matter consists of tiny indivisible and unchangeable atoms which cannot be created, destroyed or transformed into other atoms.
2. The atoms of each element have identical mass and properties. Dalton suggested that all atoms of the same element have identical weights – every atom of an element is identical to every other atom of that element.
3. Atoms of different elements can be distinguished by their different atomic weights; atoms of different elements have distinct properties.

Dalton assigned atomic weights to the atoms of the 20 elements that were known at the time, presenting his first list in 1803. This was a revolutionary concept for the day, and one which would play a part in the development of the periodic table of the elements later in the 19th century.

It wasn't until 1807 that the method by which he arrived at his list was published in a textbook by his acquaintance Thomas Thomson, and Dalton himself only published an account of his work in *A New System of Chemical Philosophy*, published in 1808 and 1810. Dalton's investigations of atomic weights were based on the mass ratios in which they combined, with the hydrogen atom (then, as now, the lightest element known) taken as the standard. However, he was hampered by his belief that the simplest compound between any two elements is always one atom each, so that, for example, he thought that the chemical formula for water was HO, not H_2O as we now know it to be, and in failing to consider that certain elements exist in molecular form rather than as single atoms, such as oxygen – O_2. It was known that when hydrogen reacted with oxygen, for instance, it was always in the proportions of 1 gram of hydrogen to 8 grams of oxygen but knowing the weights of each element involved says nothing about how many atoms of each

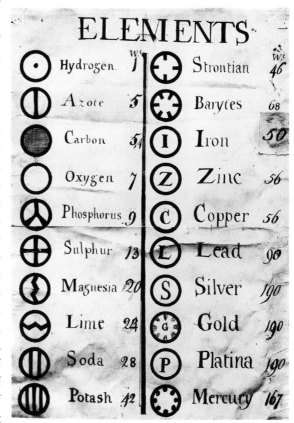

Dalton's table of the elements and their atomic weights.

there are. Dalton had to make assumptions, which turned out to be wrong, such as determining the oxygen atoms were eight times heavier than hydrogen atoms.

CHALLENGES TO DALTON

One of the first challenges to Dalton's conclusions came from a series of experiments on the volumes of reacting gases carried out by Joseph Gay-Lussac. Dalton had mainly used the weights of reacting elements to calculate atomic weights, but Gay-Lussac arrived at a law of combining volumes which stated that, at a given pressure and temperature, gases combine in simple proportions by volume. If any of the products are gaseous, they also bear a simple whole number ratio to that of any gaseous reactant. For example, 2 litres of hydrogen combine with 1 litre of oxygen to give 2 litres of water vapour. In other words, the ratio of these volumes is 2:1:2. The conclusion to be drawn from this seemed obvious, though it isn't certain if Gay-Lussac reached it: If elements combine as atoms, as Dalton asserted, and reacting gases combine in simple volume ratios, then there must be a connection between volume and the number of atoms.

Swedish chemist Jöns Jakob Berzelius (1779–1848) was a meticulous experimenter with great skill and technique in the laboratory. In 1808 he became aware of Dalton's theory and realized that, if it was right, it represented a major breakthrough in our understanding of matter. Berzelius was initially sceptical. Also in 1808, he had learned of Humphry Davy's discovery of the previously unknown metals potassium and sodium, which had been extracted by passing an electric current through potash (potassium hydroxide) and soda (sodium hydroxide) respectively. Repeating Davy's experiments for himself, and performing further ones of his own, convinced Berzelius that chemical compounds were held together by electrical attraction between the elements. If this were so, then, as he saw it, Dalton's atomic theory was 'attended with great difficulties'.

Berzelius thought that if chemical combination resulted from an attraction between oppositely charged atoms, then two-element compounds should be of the type A + B, A + 2B, A + 3B and so on, with the similarly charged B atoms distributed around the oppositely-charged A atom, and kept apart from each other. However, experiment appeared to show that some compounds had atomic compositions of the 2A + 3B or 3A + 4B type, which Berzelius thought ought to be unstable, though he eventually accepted their existence. This 'electrochemical dualist' approach was also a block to understanding that elements such as oxygen existed in molecular form.

Through his own painstaking experiments and his interpretation of the work of others such as Gay-Lussac, Berzelius produced the most reliable table of atomic weights then available. Examining Gay-Lussac's findings, Berzelius interpreted them as showing that equal volumes of gases, under identical conditions of temperature and pressure, have the same number of atoms. Taking this together with Dalton's proposal that simple whole numbers of atoms join together to give compounds, then applied to the reaction of hydrogen and oxygen to give water, for

Berzelius made an analysis of over 2000 compounds.

example, we see that two volumes of hydrogen plus one volume of oxygen yields two volumes of water, so it follows that $2n$ particles of hydrogen plus n particles of oxygen gives n particles of water. If n was equal to 100, then two volumes of hydrogen contain 200 particles of hydrogen, and one volume of oxygen contains 100 particles of oxygen. Thus, according to Berzelius, the number of atoms of hydrogen and oxygen in water would be in the ratio of 2:1; the formula of water would thus be H_2O (and not HO, as assumed by Dalton), and every atom of oxygen would be 16 times heavier than a hydrogen atom, not eight times as asserted by Dalton.

AVOGADRO AND CANNIZZARO

In 1811 Italian chemist Amedeo Avogadro (1776–1856), replaced the word 'atom' with 'molecule' in Berzelius' proposal that at a given pressure and temperature, all gases contain the same number of atoms. He appeared to be making a subtle distinction, but it was actually a proposal with profound consequences for the understanding of elements and compounds and the bringing together of Gay-Lussac's findings and Dalton's atomic theory.

It followed from Avogadro's hypothesis that the relative molecular weights of any two gases will be in the same ratio as the densities of those gases under the same conditions of temperature and pressure. Avogadro also made the assertion that simple gases, rather than being made up of single atoms, were instead formed from compound molecules of two or more linked atoms. (At this time the words 'atom' and 'molecule' were used more or less interchangeably. Avogadro referred to an 'elementary molecule', what today we would call an atom; essentially, what he was doing was defining a molecule as the smallest part of a substance.) Avogadro reasoned that Gay-Lussac's finding that a volume of water vapour was twice that of the oxygen used to form it could be explained if it was assumed that the oxygen molecule had split in two in the course of forming the water vapour. Avogadro's proposal meant that combinations by volumes in ratios of small whole numbers corresponds to combinations by particles in ratios of small whole numbers, thus reconciling Dalton's law of multiple proportions with Gay-Lussac's law of combining volumes.

It would be several years before Avogadro's proposals were accepted. The main reason for this was the continuing belief in the indivisibility of the atom and the impossibility that two atoms of the same element should combine together (oxygen was still believed to be a single

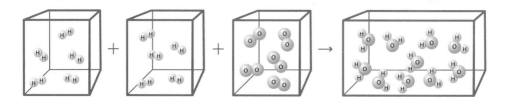

Gay-Lussac and Avogadro showed how elements combine in small whole number volume ratios.

Stanislao Cannizzaro was an indefatigable champion of Avogadro's ideas.

atom rather than a diatomic molecule). One of the main champions of Avogadro's hypothesis was Italian chemist Stanislao Cannizzaro (1826–1910).

By the middle of the 19th century, a number of competing ideas on how best to calculate atomic weights and molecular formulas were causing confusion in scientific circles, while evidence was mounting from experiments that supported Avogadro. In 1858, Cannizzaro set out his belief that a complete return to the ideas of Avogadro could be used as a basis for constructing a theoretical underpinning for almost all of the available evidence gathered from experiment. Any anomalies, he argued, could easily be accepted as minor exceptions to the general rules. In summary, declared Cannizzaro, 'the conclusions drawn from [Avogadro] are invariably in accordance with all physical and chemical laws hitherto discovered.' In September 1860, at an international chemical congress held in the German town of Karlsruhe, Cannizzaro made a compelling case for his ideas, which were soon widely adopted.

Karlsruhe, scene of the international chemical congress of 1860.

Chapter 6

ELECTRIFYING EXPERIENCES

Electrifying Experiences

TIMELINE OF DISCOVERIES ABOUT ELECTRICITY

Timeline	
6TH CENTURY BC	Thales describes amber's ability to attract light objects, such as feathers.
16TH CENTURY	William Gilbert compiles a list of materials that can be electrified by rubbing.
17TH CENTURY	Otto von Guericke invents the first electrical machine, the sulphur sphere.
1729	Stephen Gray discovers that electricity flows and is conducted by some substances but not by others.
1733	Charles Francois de Cisternay du Fay theorizes that electricity is formed of two distinct fluids which attract and repel like magnets.
1745	Pieter van Musschenbroek invents the Leyden jar, a device for storing electricity.
1752	Benjamin Franklin conducts his kite experiment, to demonstrate that electricity was the same phenomenon as seen in lightning.
1800	Alessandro Volta demonstrates the voltaic pile, an early type of battery, to the Royal Society.
1807	Humphry Davy theorizes that the forces holding matter together are electrical in nature.

Humans have been aware of electrical phenomena throughout history. Lightning strikes may well have provided early hominids with one of their first and most useful tools, fire. Early civilizations believed that lightning bolts were flung across the sky by the gods. On a smaller scale, there is no doubt that the phenomenon of static electricity causing rubbed amber to attract light materials to it was observed and wondered at. But understanding of, still less control of, electricity would be a long time coming.

Amber is the fossilized resin of a now extinct coniferous tree, almost all of which has been found in the Baltic region of Northern Europe. Because of its warm yellow colour and attractive appearance it has been greatly prized for jewellery. Amber was called 'elektron', by the Greeks, from which our word electricity originates. One of the first to write about its ability to attract

Amber is prized for its appearance as well as its interesting electrical properties.

light objects, but almost certainly not the discoverer of the effect, was the Greek philosopher Thales in the 6th century BC. The Roman writer Pliny described how spinners in Syria placed amber on the ends of their spindles, calling it 'the clutcher'. Possibly the amber became charged as the wheel spun and attracted stray bits of wool and chaff to itself.

Very little headway was made in understanding the reasons for the properties exhibited by amber for over 1,500 years until William Gilbert began his researches into magnetism (see page 21).

Gilbert also carried out many investigations into electrical phenomena, using an instrument called an electroscope, or 'versorium' as Gilbert termed it (from the Latin meaning 'to turn about'). This was a light metal compass-like (but not magnetized) needle, balanced on a pinhead at the midpoint and was a sensitive detector of electric properties. He compiled a list of over 20 materials, including glass, sapphire and sealing wax, in addition to amber that could be electrified by rubbing. He also noted that, unlike a magnetized object, an electrified object had no poles, and could be blocked by a sheet of paper. Gilbert referred to the attractive powers of amber and the like as 'electricus'.

Otto von Guericke (1602–86) was the Burgomaster (mayor) of the town of Magdeburg, Germany. As well as carrying out the experiments with vacuums for which he is best known, he also pursued his belief

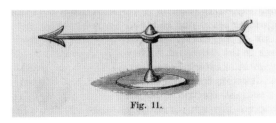

Fig. 11.

Gilbert's versorium.

that the force of gravity was electrical in nature. Similar to the magnetic earth model (the *terrella*) that Gilbert had made (see page 22), von Guericke made an electric one. This consisted of a sphere of sulphur, which he described as being about the size of a child's head, with a wooden rod through the middle. The ends of the rod

Otto von Guericke performed electrical experiments with a highly charged globe of sulphur.

were placed on supports allowing the sphere to be rotated easily. Rotating and rubbing the sphere electrified it, so that it attracted chaff, feathers and similar small objects. Von Guericke discovered that a feather that had touched the globe was repelled by it. By lifting the globe up by the rod through its middle he was able to pursue a charged feather around the room, keeping it aloft by repulsion. He also observed the phenomenon of electrical conduction, noting that a thread attached to the globe would show electrical attraction at its far end.

Amusing though the sight of von Guericke chasing feathers may have been, more important was the fact that, with his rotating sphere, he had invented the first electrical machine. The rotating sulphur sphere soon became the standard way of producing electricity and remained so for over a century. Eventually the sulphur ball was replaced by large glass cylinders or spheres supported on wooden frames, and with a piece of leather or other material to do the rubbing.

GO WITH THE FLOW

The discovery that electricity could flow is credited to English chemist Stephen Gray (1666–1736). He discovered that he could electrify corks stuck in the ends of glass tubes by rubbing the tube and that he could transmit the electrical effect to other objects by direct connection. Using string, he could charge an object over 15 metres from the rubbed tube, and brass wire would transmit the charge even more efficiently. Other substances, such as silk, wouldn't transmit the charge at all. Gray was the first to begin to divide substances into electrical conductors and non-conductors.

In a famous public demonstration in 1730, Gray suspended an eight-year-old boy from the ceiling using silk threads which acted as insulators. He then charged the boy before

Stephen Gray's experiments demonstrated that electricity could flow through some materials.

proceeding to reveal to the audience some of the effects of electricity. For example, Gray held a book near the boy's hand and asked him to turn the pages without actually touching them. When the boy held out his hand, the nearest page of the book floated up towards him, attracted by the electric charge. Next, Gray called for a volunteer who would be surprised to receive an electric shock when a spark jumped from the boy's hand to his.

At this time, scientists believed that electricity was a type of invisible fluid. In 1733, French physicist Charles François de Cisternay du Fay (1698–1739) formulated a theory of electricity that assumed the existence of two distinct electrical fluids. He had observed that objects that had been charged sometimes attracted and sometimes repelled each other. He explained this by proposing that there were two different kinds of electricity – vitreous electricity (from the Latin for 'glass') was produced by rubbing glass and some other materials, such as gemstones; resinous electricity (from the Latin for 'resin') was produced by rubbing materials such as amber.

Du Fay argued that, in a similar way to the behaviour of magnets, objects charged with the same kind of electricity repel each other, whereas those that are differently charged attract each other. Objects

Charles du Fay believed electricity was made up of two distinct fluids.

that are not electrified were assumed to have equal amounts of these fluids, which neutralize each other. Rubbing an object removed one of the fluids, leaving an excess of the other.

STORING UP TROUBLE

By the middle of the 18th century the early electricity generators had improved to the point that they were becoming dangerous. In 1745, Dutch physicist Pieter van Musschenbroek (1692–1761) invented a device for storing the electricity produced. Named after the town in which he lived, the Leyden jar was a glass jar filled with water with metal foil lining the inside and around the outside. Curious to see if electricity was indeed being stored, van Musschenbroek touched the inside and outside of the jar simultaneously, at which point he thought 'it was all up with me' as the electric charge coursed through him.

Today, we might think of the Leyden jar as an early form of capacitor. A typical design has a metal rod emerging from the cap of the jar, connected to the inner lining by a chain. The rod is charged with electricity by touching it to a charged-up glass or sulphur sphere, for example. Several jars could be connected in parallel to increase the amount of charge stored.

Van Musschenbroek steadfastly refused ever to repeat his shocking experience again but others were keen to try out the experiment for themselves. It was discovered

A Leyden jar could store literally shocking amounts of electricity.

that people could be shocked in pairs, and even in chains. French academician Louis-Guillaume Le Monnier (1717–99), who shocked 140 courtiers in the presence of King Louis XV, wrote: 'It is singular to see the multitude of different gestures, and to hear the instantaneous exclamations of those surprised by the shock.' Le Monnier also made some important new findings, including that conducting bodies become charged with electricity as a function of their surface area, not of their mass, and that water is one of the best conductors of electricity. He succeeded in sending electricity from a Leyden jar through a wire approximately 1,850 metres in length, concluding that the charge passed through the wire instantaneously.

Pieter van Musschenbroek, who advised others against repeating his painful experiment.

GO FLY A KITE

In 1747, Benjamin Franklin (1706–90) declared that du Fay's idea that there were two types of electrical fluid was unnecessarily complicated and that all the observed phenomena could be accounted for by just one fluid. According to Franklin's theory, a positive charge represents an excess of this fluid, while a negative charge is a deficit. He suggested for example that a body with an excess of electrical fluid will attract one with a deficiency and, perhaps less plausibly, that two objects with a deficiency would repel each other. Franklin's theory also explained the conservation of electric charge – when a positive charge was created, an equal negative charge was also created.

Franklin had spent the summer of 1747 conducting a series of experiments with electricity. Rather than refer to vitreous and resinous electricity he began to refer to positive and negative electricity instead. Franklin described the concept of an electrical battery in a letter to fellow scientist Peter Collinson in the spring of 1749, but he wasn't sure what practical purpose it might serve. Later the same year, he set out what he believed were similarities between a spark of electricity and

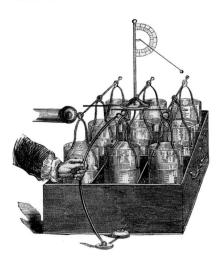

Benjamin Franklin's electrostatic battery formed from a series of Leyden jars.

lightning, such as the colour of the light, the crackling noise it made and the crooked direction it took.

Observing that a sharp iron needle would conduct electricity away from a charged metal sphere, Franklin theorized that lightning strikes might be preventable by using an iron rod connected to earth to empty static from a cloud. He came up with the idea of the lightning rod about 2 to 3 metres in length that was sharpened to a point at the end. He wrote, 'the electrical fire would, I think, be drawn out of a cloud silently, before it could come near enough to strike...'

Political lightning

The lightning rod would become an unexpectedly political statement. English scientists favoured blunt-tipped rods, reasoning that sharp ones attracted lightning and blunt rods were less likely to be struck. King George III consequently had blunt lightning rods installed. When it came to fitting buildings in the American colonies with lightning rods, favouring a pointed lightning rod became a way of expressing support for Franklin's idea for protecting buildings and the rejection of the theory supported by the King.

Franklin's most enduring contribution to the history of electricity involves his iconic kite-in-a-thunderstorm experiment. One day in June 1752, Franklin went out into a field with a kite, a key and a Leyden jar. The aim was, as Joseph Priestley later wrote: 'To demonstrate in the completest manner possible, the sameness of the electric fluid with the matter of lightning.'

Franklin constructed a simple kite from silk with a wire fixed to the kite to act as a lightning rod. A hemp string, which would conduct electricity, was attached to the bottom of the kite, and a silk string, which would not, was attached to that for Franklin to hold. The metal key was also attached to the hemp string. Noticing loose threads of the hemp string standing up, Franklin moved a finger near the key and, as Priestley later wrote, 'perceived a very evident electric spark'. He was able to collect electricity in his Leyden jar.

It is almost certain that Franklin's kite was not actually struck by lightning. Had it been he would almost certainly have been killed, and in fact the next two people who tried to replicate his experiment died in the attempt. It is more likely that the kite picked up ambient electrical energy from the storm. As it was, Franklin wasn't the first to demonstrate the electrical nature of lightning. Thomas-François Dalibard had accomplished it a month earlier in northern France.

TWITCHING FROGS

Experiments by Franklin and others had contributed greatly to science's understanding of electricity, but further progress would need a steady and reliable source of current to really study and open up its secrets.

A somewhat fanciful depiction of Franklin's kite-flying experiment.

Alessandro Volta – an example of his famous pile can be seen behind him.

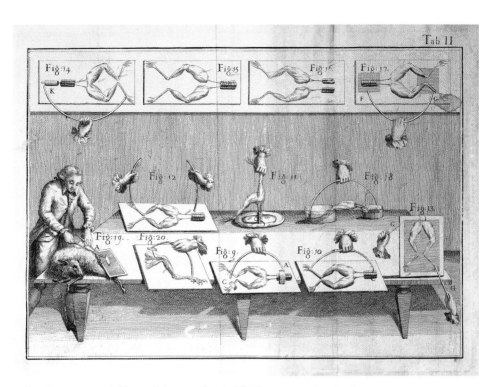

Galvani's experiments led him to believe an electrical fluid was present in animals.

Italian Alessandro Volta (1745–1827) was a self-taught physicist who learned through correspondence with scientists of the day and by conducting his own experiments in a friend's laboratory. Such was the quality of his work that he was appointed professor of physics at the University of Pavia in 1778. Electricity held a particular fascination for Volta. His greatest contribution to our knowledge of the subject came about through a dispute with another scientist.

Luigi Galvani (1737–98) was a fellow Italian physicist, physician and biologist. Around 1780 Galvani had begun investigations into the effects of electricity on dissected frogs, noting that he could make the dead frog's legs kick by applying a spark from a static electricity generator to the frog's spinal cord. Interestingly, he also noticed that muscle contractions could happen without the application of electricity. When he used his iron scalpel to dissect the leg of a frog fixed to a copper hook, the leg twitched. Galvani decided that there was a type of electrical fluid in the frog's body, naming it animal electricity. In public demonstrations, Galvani would show rapt audiences dozens of frogs' legs twitching away when they were hung from copper hooks on an iron wire.

Hearing of Galvani's discovery, Volta was intrigued. He didn't believe that the frogs were producing electricity, rather he suspected that the answer lay with the metals. He thought that the contact between the different metals was what was generating the electricity. There was no instrument available at the time that could detect such a weak output so Volta relied on the sensitivity of his own tongue, popping various combinations of metal into his mouth to see what happened.

Finding zinc and copper to give the best results, Volta built a vertical pile of alternating zinc and copper discs, separated by circles of cloth soaked in brine. When he connected a wire to each end of the pile he found that he had created a flow of electricity. He determined this was happening by touching the wires to his dampened fingers, noting that he felt a 'small prickling or slight shock', which gradually increased in force the more discs he added to the pile. For the first time, it became possible to produce a steady supply of electrical power. Up to then, the only sources of electricity had been lightning, which was impossible to control, and electrostatic devices that could generate powerful sparks, but not a sustained electrical current. Volta's pile did, and the experiments it made possible were soon firing the imaginations of scientists across the world.

Volta reported his invention to the Royal Society in London in 1800 and within weeks the voltaic pile, as it came to be known, was being put into use in laboratories. Scottish chemist William Cruickshank (1745–1800) designed a horizontal version of the pile, with the zinc and copper plates housed in an insulated wooden box, that became very popular.

DAVY'S DISCOVERIES

Humphry Davy (1778–1829) was one of the greatest scientists of his time. He made his name through his study of gases, including his investigations into the effects of nitrous oxide, or laughing gas, and had a reputation as a fearless researcher, willing to put his own life at risk. According to his biographer, Richard Holmes, 'He tests everything on himself... It's amazingly reckless... On a number of occasions, he does nearly kill himself.'

Davy was offered a position as lecturer, and later as professor of chemistry, at the Royal Institution in London in 1801. His lectures were hugely popular, drawing hundreds of people and making him one of the first scientists to be a recognized public figure.

Electromotive force

Volta believed that a type of force was causing the electric charge to move through his circuit, calling it the 'electromotive force'. The stronger the electromotive force (emf), the stronger the current that flowed. Today, we understand that the emf is not a force at all, but rather a potential difference resulting from, in this case chemical energy, that allows the current to flow. The unit of emf, the volt, is named in honour of Volta.

Danse macabre

In 1803, Giovanni Aldini, a nephew of Galvani put a somewhat macabre spin on his uncle's experiments. Having obtained the recently hanged body of convicted murderer George Foster, Aldini applied an electric current from a voltaic pile to the corpse. According to the report in *The Newgate Calender* '... the jaws of the deceased criminal began to quiver, and the adjoining muscles were horribly contorted, and one eye was actually opened... the right hand was raised and clenched, and the legs and thighs were set in motion.' To witnesses, it seemed like the dead body was being brought back to life.

Volta's invention soon came to Davy's attention, and he was quick to assemble a voltaic pile in his well-appointed laboratory. Davy postulated that the pile's source of electricity came from chemical reactions and, experimenting with different types of metal, established that one of the metals in the battery was always oxidized. He showed that an electric cell could

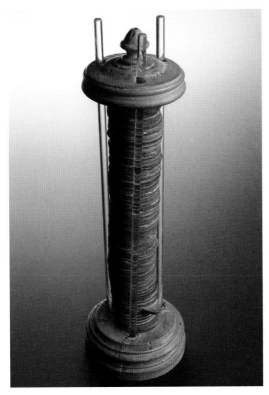

be made using just a single metal and two fluids, provided one of the fluids was capable of oxidizing one surface of the metal.

William Nicholson (1753–1815) had been one of the first in England to make his own voltaic pile and had used it to split water into its constituent parts of hydrogen and oxygen. Davy was determined to use electricity to break apart other substances and see what might be revealed.

In 1807, after several attempts, Davy succeeded in decomposing caustic potash (potassium hydroxide), finding that he had produced a metallic element that had never been seen before: potassium. Buoyed by his success, he passed the current through caustic soda (sodium hydroxide) this time bringing to light another new element: sodium.

Davy had shown how electricity could be a powerful new tool for investigating the make up of matter.

Volta's pile demonstrated that electricity flowed when different metals were brought into contact.

Over the course of the following year, Davy's battery revealed four more elements: barium, calcium, magnesium and strontium. 'Nothing,' he observed, 'promotes the advancement of science so much as a new instrument.'

Since electricity could be used to break substances apart, Davy concluded that the forces holding them together must themselves be electrical in nature. His work influenced that of Jöns Jacob Berzelius and his attempts to formulate an atomic theory of matter (see page 68). In a paper to the Royal Society, Davy wrote: 'Matter may ultimately be found to be the same in essence, differing only in the arrangement of its particles.'

The next major steps towards an understanding of electricity would be taken by a man with no formal education who started out as Davy's assistant at the Royal Institution in 1813 as a chemical assistant helping out with a variety of experiments. Within a few years he would be playing a part in linking the twin phenomena of magnetism and electricity into the new world of electromagnetism.

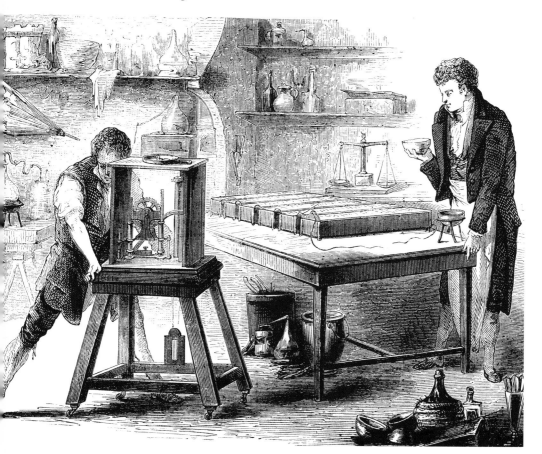

Humphry Davy experimenting with a voltaic battery.

Chapter 7
ELECTROMAGNETISM

Electromagnetism

TIMELINE OF ELECTROMAGNETISM

Timeline	
1780	Augustin de Coulomb demonstrates there is no interaction between a static electric charge and a magnet.
1820	Hans Christian Oersted observes a compass needle being defected by an electric current.
1821	Michael Faraday invents the electric motor.
1824	The first electromagnet is constructed by William Sturgeon.
1826	André-Marie Ampère proposes the existence of an electrodynamic molecule, a similar concept to the modern electron.
1831	Faraday and Joseph Henry independently discover electromagnetic induction.
1832	Faraday develops the idea of the electromagnetic force field.
1865	James Clerk Maxwell describes electric and magnetic phenomena in four equations and introduces the concept of the electromagnetic wave.

The discovery that electricity and magnetism are intimately linked – that electricity produces magnetism and magnetism generates electricity – was a groundbreaking insight.

Up until 1820, most scientists would have agreed that magnetism and electricity, while similar in some ways, were fundamentally different forces. French physicist Charles Augustin de Coulomb (1736–1806) had demonstrated in 1780, for instance, that there was no interaction between a magnet and a static electric charge. There were hints, however, that there might be some connection. Sailors had reported that sometimes following a lightning strike on their ship's mast the polarity of its compass would be reversed. Danish physicist Hans Christian Oersted (1777–1851) was one who believed there must be a link between the two forces.

Hans Christian Oersted discovered a relationship between electricity and magnetism.

On 21 April 1820, Oersted observed a phenomenon that appeared to confirm his suspicions. While setting up his apparatus for a lecture demonstration he noticed that a compass needle brought close to a wire, through which an electric current was flowing, was deflected away from its normal north-pointing direction. The effect was slight, but to Oersted it was significant. Further experiments confirmed his discovery. He tried various types of wires, all of which resulted in the needle being deflected, and noted that the effect couldn't be shielded by placing wood or glass between the compass and the wire. He tried various orientations of compass and wire and discovered that the current was producing a circular magnetic effect around it. Reversing the current deflected the needle in the opposite direction. He was certain that the electric current was producing a magnetic field (he called it 'an electric conflict').

Oersted published his results on 21 July 1820 in a paper entitled *Experiments on the Effect of a Current of Electricity on the Magnetic Needle* and caused a sensation among the scientific community. In September 1820, François Arago demonstrated the electromagnetic effect at the French Academy in Paris. One of those present was André-Marie Ampère (1775–1836). Ampère was determined to understand just why an electric current produced a magnetic effect.

Ampère began by replicating Oersted's experiments for himself and was soon making discoveries of his own. Ampère set up his experiment using countermagnets to cancel out the effect of the earth's magnetic field, enabling him to detect the effect Oersted had observed with more reliability and sensitivity. He found, for example, that if electric currents flow in the same direction through two nearby parallel wires, the wires attract one another; if the electric currents are flowing in opposite directions the wires repel one another. Without using magnets, Ampère had created a magnetic force. He deduced what is now called the righthand rule: if the thumb of the right hand is placed along the wire, pointing in the direction of the current flow, the fingers of the right hand curl around the wire in the direction of the magnetic force.

To explain the relationship between electricity and magnetism, Ampère proposed the existence of a new particle responsible for both of these phenomena, which he called the 'electrodynamic molecule', and which we now know to be the electron. Ampère believed, correctly, that huge numbers of these electrodynamic molecules were moving in electric conductors and suggested that they moved in tiny closed circuits inside a magnetic substance. He also developed a mathematical equation linking the size of the magnetic field to the electric current that produced it.

Ampère's most important publication on electricity and magnetism was published in 1826. It is called *Memoir on the Mathematical Theory of Electrodynamic Phenomena, Uniquely Deduced from Experience*. His theory became fundamental for 19th century developments in electricity and magnetism.

André-Marie Ampère laid the groundwork for the study of electromagnetism.

MICHAEL FARADAY

Michael Faraday (1791–1867) built his own voltaic pile at the age of 19, having seen one in operation at the City Philosophical Society in London. After attending a series of lectures on chemistry given by Humphry Davy in 1812 he sent his lecture notes to Davy with a request to be taken on as his assistant. A few months later Davy arranged for Faraday to join him at the Royal Institution.

Between 1813 and 1820 Faraday assisted Davy with his experiments while at the same time making the acquaintance of some of the leading scientists of the day. In 1821, Faraday was invited to write an article for the *Annals of Philosophy*, summarizing what was known so far of the nascent field of electromagnetism. In coming to terms with

Michael Faraday's accomplishments included the invention of the electric motor and the discovery of electromagnetic induction.

Oersted's discovery Faraday began a research trajectory that he would follow for the rest of his life.

Examining Oersted's observation regarding the circular magnetic forces around a wire, Davy performed a simple experiment in which he passed a wire up through a piece of paper, held at right angles to the wire, on which were scattered iron filings. Sure enough, when the current was turned on the filings formed circular patterns around the wire. Faraday, Davy and physicist William Hyde Wollaston (1766–1828) discussed whether there was any potential for harnessing this magnetic field to produce motion. Wollaston's attempts to do so failed. Faraday carried out further experiments of his own, observing the lines of force around permanent magnets.

Faraday had the idea that if a wire carrying a current exerts a magnetic force, the magnetic force should exert an opposite force on the wire. He set up an ingenious experiment

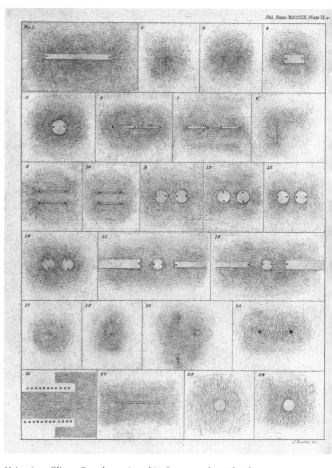

Using iron filings, Faraday painstakingly mapped out the shape of the magnetic field.

in which he prepared a beaker of mercury with a bar magnet fixed vertically at the centre. A wire that was free to move was suspended from a metal arm so that it dipped into the mercury, which completed the electrical circuit. On Christmas Day, 1821, he connected his experiment and was delighted to find the wire in the beaker rotating about the magnet as he expected. The current flowing through the wire produced a magnetic field which interacted with the field from the bar magnet. Faraday had demonstrated the first continuous motion created by electromagnetism and had pointed the way towards the electric motor. Faraday neglected to acknowledge the influence of Davy and Wollaston and there was subsequent ill feeling between them.

One way of defining an electric motor is as a device that converts electrical energy into kinetic energy. By that definition, Faraday had invented the first electric motor with his

Christmas experiment in 1821. In 1822, Peter Barlow replaced the dangling wire with a spoked wheel that dipped into a trough of mercury. When the current was applied the wheel began to rotate, perhaps hinting more than Faraday's spinning wire at the uses the newly-discovered phenomenon might be put to.

ELECTROMAGNETIC INDUCTION

Much of Faraday's time in the 1820s was spent working on a project for the Admiralty so he was kept from returning to his electrical researches until 1831. That year, Faraday made a discovery that paralleled the one made by Oersted a decade earlier. He found that moving a magnet near a circuit would make a current flow in the circuit. The faster the magnet moved, the stronger the current generated. If the wire was coiled so that more of it was exposed to the magnetic field it again made the current stronger. Faraday tried another idea. He took an iron ring and wound two coils of wire on opposite sides. The first coil was connected to a battery, while the second coil was simply connected to itself. Faraday reasoned that when the current was switched on in the first coil, a magnetic field would be created in the iron, which in turn would cause a current to flow in the second coil. At first, the experiment was a disappointment – Faraday detected no current in the second coil. But he did see something curious.

When the first coil was switched on, there was a flicker of movement from a compass needle that Faraday had hoped would detect an electric current in the second coil. It then returned to its original position. When the coil was switched off, the needle again flicked briefly, this time in the opposite direction. Faraday quickly realized what was

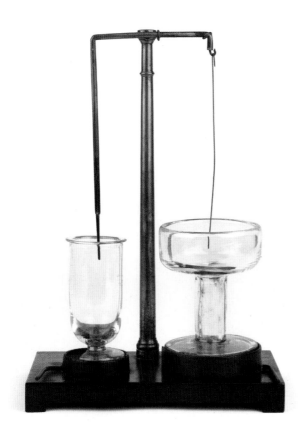

Faraday discovered how to convert electric current into motion.

The electromagnet

In the course of his experiments Ampère had placed an iron rod inside a coil of wire and found that it behaved like a permanent magnet when the current was turned on. In 1824, William Sturgeon, a British Royal Artillery officer, constructed the first U-shaped electromagnet by winding a few turns of bare copper wire around an iron horseshoe. In 1829, Joseph Henry (1797–1878), improved the idea by using many turns of insulated copper wire around a thicker U-shaped iron core, eventually building a magnet that could lift a tonne of iron. Henry is also credited as the co-discoverer, with Michael Faraday, of the principle of magnetic induction – the generation of an electric current by a changing magnetic field. Less than a decade after Oersted's discovery, science was finding practical uses for the powerful magnetic forces that could be generated.

happening – changing magnetic fields generated electric currents just as a moving magnet did. By turning the current in the first coil on and off rapidly he could make a current flow in the second coil. Faraday had discovered electromagnetic induction. He published his results in 1832, ahead of Joseph Henry, who had also recognized the phenomenon during the course of his experiments with electromagnets. In Henry's honour the SI unit of inductance is called the henry.

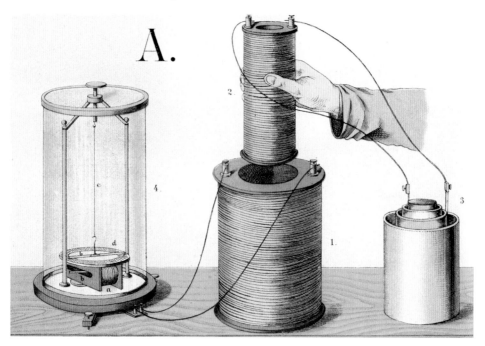

Faraday's induction demonstration; moving the small coil through the large one creates a changing magnetic field.

That magnetism and electricity were related on a fundamental level could no longer be in doubt. It was a world-changing discovery, one that lies behind all of the electricity generated in the world's power stations and which drives all of the countless electric motors we use.

Faraday was also convinced that there was a link between electricity, magnetism and light. In 1845, he carried out an experiment in the basement of the Royal Institution and found that he could affect the polarization of a beam of light using an electromagnet, indicating that light did indeed have electromagnetic properties. Faraday wrote in his notebook: 'I have at last succeeded in ... magnetizing a ray of light'. This demonstration of what is now called the Faraday effect was an important stepping stone in the development of the field theory of electromagnetism.

Faraday wanted to explain how a magnet could induce an electric current in a wire without coming into physical contact with it, and how an electric current could make a compass needle move. To do so, he came up with the idea of an electromagnetic field. He saw this as lines of force, which he called 'flux lines', stretching invisibly through all of space – the lines that were made visible in the patterns formed by iron filings scattered on a sheet above a magnet. According to Faraday's field theory, the magnetic lines of force were concentrated around the magnet, rather than the magnet actually creating the field. The magnet was not the centre of the magnetic force but rather it concentrated the force through itself. The magnetic force wasn't in the magnet but in a magnetic field in the space surrounding it.

The law of induction

Faraday's law of induction states that the induced voltage in a circuit is proportional to the rate of change over time of the magnetic flux, or total magnetic field, passing through that circuit. In other words, the faster the magnetic field changes, the greater will be the voltage in the circuit. The direction of the change in the magnetic field determines the direction of the current. The greater the number of loops in the circuit the greater the voltage. The induced voltage in a coil with four loops will be twice that with two loops. A higher voltage can also be obtained by spinning a generator faster.

BRINGING IT ALL TOGETHER

Scottish physicist James Clerk Maxwell is acknowledged as being one of the finest scientists who have ever lived. Some 20 years after Faraday proposed his field theory, Maxwell took up the idea and set out to express Faraday's ideas mathematically. Albert Einstein would later describe Maxwell's work on electromagnetism as 'the most profound and most fruitful that physics has experienced since Newton'.

In just four short equations, Maxwell succeeded in describing all of the electric and magnetic phenomena that had been observed and recorded by Faraday and other researchers.

Maxwell's equations described all of the different aspects and behaviour of both forces and provided accurate predictions for future experiments. Writing in 1940, Einstein declared 'it took physicists some decades to grasp the full significance of Maxwell's discovery, so bold was the leap that his genius forced upon the conceptions of his fellow workers'.

Maxwell's four equations describe the electric and magnetic fields associated with electric charges and currents, and how those fields change over time. They brought together decades of experimental observations of electric and magnetic phenomena by people such as Michael Faraday and André Ampère. Maxwell established for the first time that varying electric and magnetic fields could propagate indefinitely through space in the form of electromagnetic waves as one generated the other.

An electromagnetic wave can be imagined as being like two waves travelling in the same direction but at right angles to each other. One of these waves is an oscillating magnetic field, the other is an oscillating electric field. The two fields keep in step with each other as the wave travels along. Maxwell's proposals showed that electricity and magnetism were always bound together and that it was impossible to have one without the other. Using his equations, Maxwell calculated the velocity of an electromagnetic wave as 299,792,458 metres per second, a value that agreed with the experimental value for the velocity of light. Maxwell thought this couldn't possibly be a coincidence and declared that light itself was an electromagnetic wave. In Maxwell's equations, the speed of electromagnetic waves is a constant defined by the properties of the vacuum of space through which the waves move. The nature of the universe and the behaviour of electric and magnetic fields dictate the speed with which electromagnetic waves are propagated.

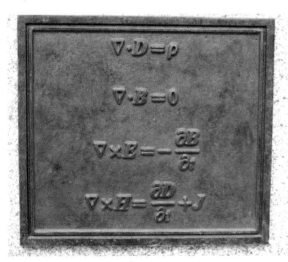

James Clerk Maxwell's four equations.

He accurately predicted that there should be a whole range, or spectrum, of electromagnetic waves. Infrared and ultraviolet light, invisible to human eyes, had already been discovered at either end of the visible spectrum and scientists had demonstrated that they had the same wavelike properties as visible light. After Maxwell's death, the discovery of long wavelength radio waves and very short wavelength X-rays and gamma rays extended the spectrum further.

Chapter 8

THERMODYNAMICS

Thermodynamics

TIMELINE OF THERMODYNAMICS

Timeline	
1698	Invention of the steam engine.
1761	Joseph Black discovers latent heat, which helped establish the theory of thermal equilibrium, the Zeroth Law of Thermodynamics.
1787	Antoine Lavoisier proposes the existence of caloric, the substance of heat.
1798	Count Rumford determines that heat is generated by friction.
1811	Joseph Fourier publishes a theory of heat conduction.
1820	Nicolas Carnot lays the foundations of thermodynamics.
1840	Julius Robert von Mayer proposes a 'conservation law' for energy.
1840s	Mayer and James Joule independently arrive at the idea of the mechanical equivalent of heat.
1848	Lord Kelvin proposes an absolute scale of temperature.
1850	Rudolf Clausius establishes the Second Law of Thermodynamics – that heat always flows from a hot object to a cooler one.
1860	Gustav Kirchhoff introduces the idea of the 'black body'.
1876	Ludwig Boltzmann solves the so-called 'reversibility paradox.'
1900	Max Planck finds a way around the 'ultraviolet catastrophe' by proposing that energy is emitted in quanta.

Once reliable thermometers came into use around the beginning of the 18th century, more precise observations of temperature and heat flow became possible. Joseph Black (1728–99), a professor at the University of Edinburgh, noticed that a collection of objects at different temperatures, if brought together, will all eventually reach the same temperature, or what we now call thermal equilibrium, at which point no further heat flow takes place. Eventually this became enshrined as the Zeroth Law of Thermodynamics by British physicist Ralph Fowler in the 1930s. It states that if two objects are in thermal equilibrium with a third then they are in thermal equilibrium with each other.

Eighteenth-century scientists experimenting with electricity had speculated that it took the form of an invisible fluid. Perhaps, it was thought, heat, which seemed to flow invisibly from one place to another, was also a type of fluid. In 1787, Antoine Lavoisier, one of the founders of modern chemistry, set out a list of 33 elements, substances that chemical processes had failed to break down into simpler forms. Alongside such things as hydrogen, oxygen and sulphur, he listed *caloric*, or calorific fluid, the weightless substance of heat. Lavoisier proposed that hot bodies contain more calorific fluid than cold bodies and that the particles of calorific fluid repelled one another, causing heat to flow from hot to cold bodies. Caloric theory could also be used to explain the change from one state of matter to another. Caloric fluid flowing between the atoms of a solid weakened the attraction between them,

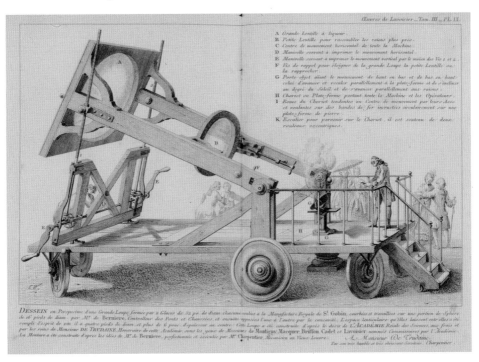

Antoine Lavoisier's solar furnace.

Count Rumford showed how the friction from boring a cannon generated more heat than could be accounted for by caloric fluid.

resulting in the solid melting into a liquid. Continued accumulation of the caloric fluid pushed the atoms still further apart, resulting in a gas, with each atom or molecule in the gas surrounded by a ball of caloric fluid.

By the turn of the century, challenges to Lavoisier's caloric hypothesis appeared, among them the observations of Benjamin Thompson, Count Rumford (1753–1814) in 1798. Born in what was then the colony of Massachusetts, Rumford was a talented engineer and inventor, although he also had a reputation as an unscrupulous self-promoter and womanizer. His lasting contribution to the physics of heat, presented to the Royal Society in 1798, happened while he was in Munich. Here he observed the heat generated by the

Joseph Fourier set out a mathematical theory of heat conduction.

boring of cannons, a task that was accomplished by turning an iron bit inside a brass cylinder. Rumford determined that the heat was generated by friction between the bit and the cylinder and not, as had previously been supposed, by caloric fluid being released from the cannon. He wrote: '... the source of the Heat generated by Friction, in these Experiments, appeared evidently to be inexhaustible... anything which any insulated body, or system of bodies, can continue to furnish without limitation, cannot possibly be a material substance'

In 1811, Joseph Fourier (1768–1830) published a mathematical theory of heat conduction. Fourier's theory proposed that the rate of heat flow between two points was proportional to the temperature difference between the points, and inversely proportional to the distance between them. Fourier did not speculate about the nature of heat, but simply considered how it appeared to behave, founding his theory on observation and experiment.

Before the Industrial Revolution, mechanical power was supplied by the natural sources of wind and water or the muscles of people and animals. The invention of the steam engine changed that. The trouble was that this new source of power was hugely inefficient, only around 3 per cent of the fuel burned was converted into useful work. Engineers tried various ways of improving the efficiency of their engines but they were hampered by the fact that they lacked insight into the key energy process involved – the transfer of heat. It was the search for this knowledge that resulted in the development of thermodynamics – the science of heat.

CARNOT'S ENGINE

The foundations of thermodynamics were laid in 1820 by a young French soldier called Nicolas Léonard Sadi Carnot (1796–1832) when he set out to tackle this problem. Carnot wanted to find a way of improving the efficiency of French steam engines, which he thought suffered in comparison to their British counterparts. Rather than look at the moving parts of the steam engine, Carnot directed his attention to the movement of heat through the engine, seeing parallels with the movement of water over a waterwheel. Just as water always flows downhill, and the waterwheel uses the falling water to do work, so Carnot imagined the steam engine using the caloric fluid 'falling' from a hot to a cold object to do work.

In the steam engine, heat is used to turn water into steam, which is directed through a pipe to a cylinder where it pushes a piston. The piston is used to do work, the steam cools down, and the vapour is expelled, leaving the piston ready for the next cycle. Carnot argued that if all friction were eliminated it was possible to imagine a reversible heat engine. He imagined an idealized heat engine in which the work output would be the same as the

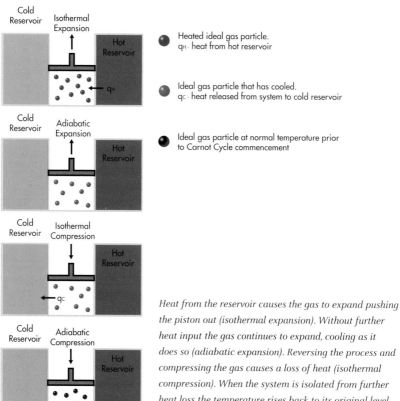

Heat from the reservoir causes the gas to expand pushing the piston out (isothermal expansion). Without further heat input the gas continues to expand, cooling as it does so (adiabatic expansion). Reversing the process and compressing the gas causes a loss of heat (isothermal compression). When the system is isolated from further heat loss the temperature rises back to its original level (adiabatic compression).

Nicolas Léonard Sadi Carnot, who laid the foundations of thermodynamics.

heat input with no energy lost in the conversion. In what is now known as a Carnot engine, gas heated in a cylinder pushes a piston as it expands. The heat supply is then cut off, but the hot gas continues to expand, cooling down as it does so. The direction of the piston is reversed, and the heat generated by the compression of the gas flows out into a heat reservoir, until a certain point is reached and the reservoir is disconnected. Then further compression heats up the gas to its original temperature, at which point the cycle begins again. After experimenting Carnot realized that it was impossible to prevent some heat being lost to the environment, but his ideas helped designers improve the efficiency of their engines.

The efficiency of a waterwheel could be calculated using straightforward Newtonian physics. The rate of flow of water and the height the water fell to the waterwheel gave the power input, and the power output could be calculated from the height that a known weight could be raised against gravity in a given time. The ratio of input to output gave the efficiency of the engine. This method couldn't be applied to steam engines, however.

THE FIRST LAW OF THERMODYNAMICS

What is energy? It's a superficially simple question that becomes increasingly vexing the more you think about it. It is often defined as 'the ability to do work', but what is it that actually delivers that ability? There is no such thing as 'pure energy', nothing that can be put in a bottle and measured. All we can really measure are the effects energy has – but not the thing itself. Energy makes everything happen, but energy itself is a mystery.

The work of scientists such as Robert von Mayer and James Joule paved the way towards what would become the First Law of Thermodynamics. In its simplest and most direct form it can be stated as: Energy can be neither created nor destroyed.

In 1840, German physician Julius Robert von Mayer (1814–78) signed on as ship's doctor for a voyage to the tropics. In the course of the journey von Mayer noticed something strange. When he was practising bloodletting on ill crewmen he was surprised to see that their venous blood was almost as bright red as their arterial blood. Back home in the cooler conditions of Germany venous blood was much darker. Mayer was familiar with Lavoisier's view that body heat is produced by a slow combustion of food and concluded that less nutrition was required to warm the body in the tropics and there were, therefore, fewer waste products in the blood. He was the first person to describe the chemical process of oxidation as the primary source of energy for life.

Mayer speculated not only about the conversion of food to heat in the body, but also the fact that the body can do work, and energy had to be provided for this too. He came to the view that heat and work are interchangeable – that the food we eat can be converted to either heat or work, but the total energy has to remain the same. He believed this principle applied not just to living things but to all systems using energy. In 1841, Mayer wrote his first scientific paper, in which he set out what he called a 'conservation law of force' (by which he meant energy) and later presented a value in numerical terms for the mechanical equivalent of heat – the amount of work required to produce a unit amount of heat.

Mayer's ideas regarding the conservation of energy attracted little attention, perhaps not helped by his tendency to express his ideas in an obscure philosophical style. At around the

The conservation of energy

'There is a fact, or if you wish, a law, governing all natural phenomena that are known to date. There is no known exception to this law... The law is called the conservation of energy. It states that there is a certain quantity, which we call energy, that does not change... it is not a description of a mechanism, or anything concrete; it is just a strange fact that we can calculate some number and when we finish watching nature go through her tricks and calculate the number again, it is the same.'

Richard Feynman, 1964

Julius Robert von Mayer.

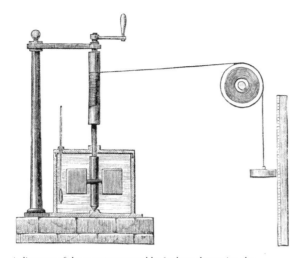

A diagram of the apparatus used by Joule to determine the mechanical equivalent of heat.

same time in England, brewer's son James Joule (1818–89) was thinking along similar lines, although he approached the problem from a different angle. Mayer was the first to set out the concept of the mechanical equivalent of heat, closely followed by Joule, but Joule was the first to back it up with firm experimental evidence. He attached a weight to one side of a pulley with a paddle wheel immersed in water on the other side. When the weight was dropped it turned the wheel, stirring the water. Joule measured the water temperature before and after rotating the wheel and found it was warmer after. Joule arrived at the conclusion Mayer had drawn, that heat and mechanical work were equivalent: a given amount of work could be transformed into a measurable and predictable amount of heat.

THE SECOND LAW

Rudolf Clausius (1822–88) is considered to be one of the founding thinkers of thermo-dynamics. In 1850, as professor of physics in the royal artillery and engineering school in Berlin, he published a paper titled 'On the Moving Force of Heat and the Laws of Heat which May Be Deduced Therefrom' setting out his ideas on heat and work, including the shortcomings he saw between Carnot's ideas and the concept of the conservation of energy. Clausius believed that heat was a result of the kinetic energy of the moving particles that made up an object. He saw that Carnot's theoretical heat engine was flawed in that the 'caloric fluid' Carnot believed flowed through it wasn't actually conserved, as less of it was dumped in the cold reservoir than was taken from the hot one, the difference being made up by the work done by the engine.

The calorific theory had been highly successful in describing heat flow through solid materials, with heat flowing from a region of high temperature to one of lower temperature, but the energy conservation law would work equally well if heat flowed from low temperature to high. Clausius therefore proposed a second law of thermodynamics which stated that: Heat always flows from a hot object to a cooler one and never the other way around. Eventually the system will reach a state of equilibrium. Once your coffee gets cold it isn't going to get

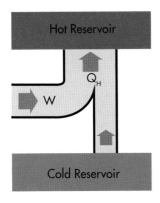 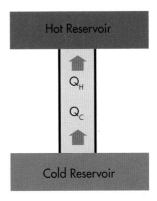

The Second Law forbids the movement of heat from a cold region to a hot one. Refrigerators need an input of energy to make this happen.

hot again. There are many ways of formulating this law. William Thomson (Lord Kelvin) (1824–1907) put it like this:

'It is impossible to devise an engine which, working in a cycle, shall produce no effect other than the extraction of heat from a reservoir and the performance of an equal amount of mechanical work.'

In 1848, Thomson's work on thermodynamics led him to propose an absolute scale of temperature based on his studies of the theory of heat, in particular the calorific theory proposed by Carnot. The Kelvin absolute temperature scale, as it became known, was precisely defined much later after the conservation of energy had become better understood.

ENTROPY AND THE ARROW OF TIME

Some physical laws, such as Newton's laws of motion, are not time dependent. Using Newton's laws, if we know how an object is moving now, we can work out how it was moving in the past, and how it will move in the future. But these laws are time reversible, and work just as well in either direction, whether we are calculating motion into the future or backwards into the past.

According to kinetic theory, heat is a measure of the motion of atoms. The more agitated the atoms, the greater the heat. Just as with any other objects in motion, collisions between individual molecules are completely reversible, yet if two gases are mixed together they will never spontaneously separate into their individual components even though, in theory, they could. Ludwig Boltzmann (1844–1906) used the kinetic theory to resolve this so-called 'reversibility paradox' in physics. He realized around 1876 that there must be many more disordered states for a system than there are ordered states; therefore random interactions will inevitably lead to greater disorder.

This arose from the second law of thermodynamics, which dictates that most natural processes are irreversible as to reverse them would involve the reversal of an energy flow, which the second law forbids. Entropy (broadly defined as a measure of disorder) and energy are similar in that an object or system may be said to have a certain 'entropy content' just as it has a certain 'energy content'. However, while the first law of thermodynamics ensures that the energy of a system is always conserved, the second law of thermodynamics ensures that the total entropy of an isolated system cannot decrease: it may (and generally does) increase.

Ludwig Boltzmann resolved the reversibility paradox.

The universe proceeds inexorably from a state of low entropy (order) to a state of high entropy (disorder), which appears to contradict the time reversible nature of Newtonian mechanics. The idea of an 'arrow of time', pointing from the past to the future, was first introduced by the astronomer Sir Arthur Eddington.

Boltzmann solved the reversibility paradox by determining that the second law was about probabilities. All of the countless atoms and molecules that make up an object are in constant random motion. There is a vanishingly small, but not absolutely impossible, chance that the molecules in a broken egg will all move in just the right direction to reassemble the egg. But the chances of this happening are just so utterly improbable that the breaking of the egg is effectively irreversible. Albert Einstein thought Boltzmann's theory was 'absolutely magnificent'. Between 1902 and 1904, Einstein himself worked on the second law of thermodynamics, developing a 'general molecular theory of heat' that would extend Boltzmann's work on gases to other materials.

A consequence of the second law and the ideas of entropy and irreversibility is that the universe, as a closed system, must eventually approach a state in which its entropy has the highest possible value. It will reach a state of equilibrium in which all forms of fuel will have been expended and all available energy will have been converted into heat, the temperature will be uniform throughout the cosmos, with no prospect of heat flowing from one place to another and therefore no possibility of any work being accomplished. In another phrase made popular in the 1930s by Sir Arthur Eddington, the universe will have entered into a final state of 'heat death'.

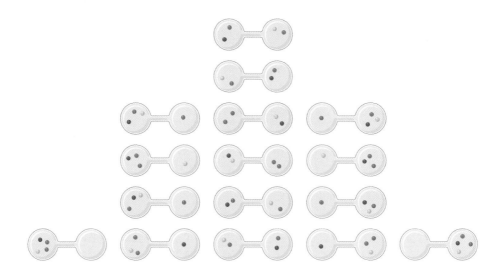

There are many more ways for the molecules of a gas to be divided between two containers than there are ordered states with all the molecules in one container.

THERMAL RADIATION

Heat energy can be transferred from one place to another in three ways: by conduction in solids, by convection in fluids and by radiation. This thermal, or heat, radiation, like radio waves, visible light and x-rays, is a form of electromagnetic radiation, the existence of which was first suggested by James Clerk Maxwell in 1865.

English astronomer William Herschel (1738–1822) had been one of the first to see a connection between heat and light. In 1800, he had measured the temperature at different points within the visible light spectrum and noticed that the temperature increased as he moved his thermometer from the violet to the red end of the spectrum. To his surprise, he discovered that his thermometer registered an increase in temperature beyond the red end of the spectrum, where no light was visible. He had discovered infrared radiation, which is invisible to the eye, but which we can detect as heat.

In 1858, Scottish physicist Balfour Stewart (1828–87) presented a paper called 'An Account of Some Experiments on Radiant Heat.' He had been investigating the abilities of different materials to absorb and emit heat and discovered that a material that tends to absorb energy at a certain wavelength tends also to emit energy at that same wavelength. Two years later, German physicist Gustav Kirchhoff (1824–87), unaware of Stewart's work, published similar conclusions. His fellow physicists found Kirchhoff's work to be more rigorous than Stewart's with the result that, although Stewart had made the discovery first, his contribution was largely forgotten.

It is possible to imagine an object that perfectly absorbs all of the electromagnetic radiation that strikes it. As no radiation is reflected from it, all of the energy that it emits depends solely on its temperature. Physicists call these hypothetical objects black bodies, a name that was coined by Kirchhoff. Kirchhoff proposed a law of thermal radiation, which stated that, for an object in thermodynamic equilibrium, the amount of radiation absorbed by the surface is equal to the amount emitted for any given temperature and wavelength. The efficiency with which an object absorbs radiation at a given wavelength is the same as the efficiency with which it emits energy at that wavelength. More concisely, absorptivity equals emissivity. Most of the energy output of a blackbody is concentrated around a peak frequency, which increases as the temperature increases. The spread of energy emitting wavelengths around the peak frequency form a distinctive shape called a blackbody curve.

In 1893, Wilhelm Wien (1864–1928) discovered the mathematical relationship between temperature change and the shape of the blackbody curve. He found that the wavelength at which the maximum amount of radiation is emitted multiplied by the temperature of the blackbody was always a constant. It meant that the peak wavelength could be calculated for any temperature and it explained why things change colour as they get hotter. As the temperature increases the peak wavelength decreases, moving from longer infrared waves to shorter blue-white and ultraviolet. By 1899, however, careful experiments were showing that Wien's predictions weren't holding true in the infrared range.

In 1900, Lord Rayleigh (1842–1919) and James Jeans (1877–1946) came up with a formula that seemed to explain what was going on at the red end of the spectrum, but they soon ran into problems of their own. According to Rayleigh and Jeans' theory, there was effectively no upper limit to the higher frequencies that would be generated by the blackbody radiation, which meant that an infinite number of highly energetic waves would be produced. This came to be known as the ultraviolet catastrophe, and it was obviously wrong, but Rayleigh's equation was based on sound physical principles and no one could explain why it didn't work.

That same year, Max Planck in Berlin was himself working on a theory of blackbody radiation, and in October 1900, he came up with an explanation for the blackbody curve that agreed with all the experimental measurements. His solution was a radical one that involved an entirely new way of looking at the world.

Max Planck, the father of quantum physics.

Lord Rayleigh (pictured) and James Jeans' attempts to explain the blackbody curve failed at higher frequencies.

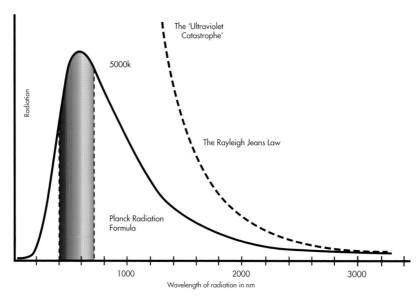

The radiation formula calculated by Planck avoided the Ultraviolet Catastrophe predicted by Rayleigh and Jeans.

Planck found that catastrophe could be averted by assuming that a blackbody emitted energy not in continuous waves, but in discrete packets, which he called quanta. The day he presented his findings to a meeting of the Deutsche Physikalische Gesellschaft (German Physical Society) in Berlin, 19 December 1900, is generally accepted as marking the birth of quantum mechanics and a new era in physics.

Absolute zero

All of the objects in the universe are exchanging electromagnetic radiation with each other all the time. This constant flow of energy from one object to another prevents anything from ever cooling to absolute zero, the theoretical minimum of temperature at which an object transmits no energy at all. Everything that has a temperature above absolute zero (equal to -273.15°C / -459.67°F) emits radiation. The hotter the object, the more energy it emits. If an object is hot enough the radiation it emits can be seen as visible light. All heated objects emit light of the same colour at the same temperature.

Chapter 9

EVERYTHING IS RELATIVE

Everything is Relative

TIMELINE OF RELATIVITY

Timeline	
1887	Albert Michelson and Edward Morley carry out an experiment using an interferometer and discover that the speed of light always remains the same.
1889–92	Hendrik Lorentz and George FitzGerald independently come up with the same theory to explain the Michelson–Morley results: the Lorentz–FitzGerald contraction.
1905	Albert Einstein sets out the theory of special relativity, including his famous equation $E=mc^2$.
1907	Hermann Minkowski develops spacetime diagrams.

The idea of relativity in physics is a fairly straightforward one. We've already come across Galileo's thoughts on observers in relative motion (see page 41). Basically, what Galileo asserted was that the laws of physics apply and remain the same for all freely moving observers, whatever their speed of motion. As Newton pointed out in his first law of motion, an inertial state, that is moving with a constant speed and direction, is the default for any object not being acted on by a force. Inertial motion is simply motion at a uniform speed in a straight line. Objects in uniform motion, moving at a constant speed and direction relative to one another are said to share an inertial frame of reference.

The idea that motion has no meaning without a frame of reference is fundamental to Einstein's theories of relativity. We can only make measurements with reference to something else. It's meaningless to say an object is massive, or fast moving, unless we stipulate what we are measuring it against. To say something is moving only has meaning if we can say what it is moving relative to. To the traveller on a train their discarded newspaper isn't moving, but to the observer by the railway track both newspaper and passenger are speeding by. The speed at which an object is perceived to be moving depends on the speed of the observer relative to that object.

Before Einstein, Newton's idea of absolute motion was generally accepted. This was the idea that an object could be said to be moving without reference to anything else; an idea that

Albert Einstein is one of the few scientists who truly transformed the way we look at the world.

required that there must also be a state of absolute rest. Either it was moving, or it wasn't. Newton wrote: 'Absolute motion is the translation of a body from one absolute place into another; and relative motion, the translation from one relative place into another.'

ON THE ELECTRODYNAMICS OF MOVING BODIES

Researchers into electromagnetism, such as Michael Faraday (see page 94), had established that an electric current is generated if a magnet is moved inside a coil of wire, and also if the magnet remains fixed and the coil is moved. It was generally assumed that there were two different mechanisms at work – one for the moving magnet producing a current and another for the moving coil producing a current. The distinction between moving magnet and moving coil depended on the view still held by most scientists that there was a state of absolute rest. Einstein asserted that it didn't matter which was moving, it was their movement relative to each other that generated the current. The idea of absolute rest was one he rejected as flawed and unnecessary.

Einstein published his ideas in a paper, his third of 1905, called On the Electrodynamics of Moving Bodies. Here he set out his 'Principle of Relativity':

'The same laws of electrodynamics and optics will be valid for all frames of reference for which the laws of mechanics hold good.'

In many ways this is was what Galileo was saying in 1632. The laws of physics are the same in all inertial frames of reference, any experiment carried out will produce results that are in accordance with those laws and no experiment can determine the motion of the observer in an inertial frame.

MICHELSON AND MORLEY

Waves need some sort of medium to carry them – sound waves travel to your ear through the medium of the air, for example. So how do electromagnetic waves travel through the vacuum of space? It was the view of 19th century scientists that light must also travel through a medium. They called it 'ether'.

As the earth moved in its orbit around the sun, the flow of the ether across the earth's surface should, it was believed, produce an 'ether wind'. A beam of light travelling through the ether should travel faster if it was moving with the wind and slower if it was going against it. American scientists Albert Michelson (1852–1931) and Edward Morley (1838–1923) carried out a series of precise experiments aimed at measuring the effects of the ether on the light that passed through it. Their crucial experiment, carried out in 1887, set out to measure the speed of light in different directions and so determine the speed of the ether relative to the earth.

To carry out the measurements, Michelson designed a device called an interferometer. This sent the beam from a light source through a half-silvered mirror which split the light

$$V = \frac{4n r n D}{\delta}$$

Michelson and Morley used interference patterns to precisely measure the velocity of light.

into two beams travelling at right angles to one another. The beams were then reflected back to the middle by two more mirrors. The beams recombined, producing an interference pattern that could be observed through an eyepiece. Any variation in the time it took for the beams to travel between the mirrors would be seen as a shift in the interference pattern. If the ether theory was correct, the speed of the light beams would change as their direction changed in relation to the direction of the earth's orbit.

Michelson and Morley discovered that it made not one bit of difference how they rotated the apparatus. Nor did it matter at what time of day they took their measurements. The speed of the light beams was always the same.

Physicists, including Michelson himself, were perplexed by this result. Did it mean that the ether didn't exist? No one doubted the reliability of Michelson and Morley's experiment, but there was reluctance to accept its conclusions. Michelson repeated the experiment, even

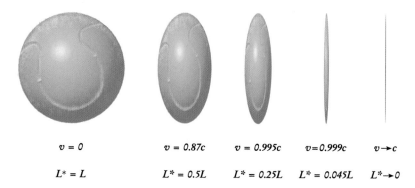

$v = 0$	$v = 0.87c$	$v = 0.995c$	$v = 0.999c$	$v \rightarrow c$
$L^* = L$	$L^* = 0.5L$	$L^* = 0.25L$	$L^* = 0.045L$	$L^* \rightarrow 0$

Einstein showed that objects contract along their direction of movement, an effect called the Lorentz–FitzGerald contraction.

trying it on a mountaintop, but the speed of light remained the same – there was not the slightest indication that the ether, if it existed, had any effect on it.

Physicists looked for ways to explain these findings that could be reconciled with their belief in the ether. One suggestion seemed little better than a fudge. Working independently of each other, Dutch physicist Hendrik Lorentz (1853–1928) and Irish

Hendrik Lorentz, who tried to explain the Michelson–Morley results.

physicist George FitzGerald (1851–1901) came up with the same solution to the problem. In 1889, FitzGerald published a short paper in which he proposed that the results of the Michelson–Morley experiment could be explained if objects reduced in length as they travelled through the ether. Lorentz put forward an almost identical proposal in 1892.

The reduction in length due to the Lorentz–FitzGerald contraction, as it came to be called, was infinitesimal, amounting to just a couple of centimetres for an object the size of the earth, but it would be enough to explain Michelson and Morley's results. What Einstein would show is that, yes, objects did shrink, but not for the reasons Lorentz and FitzGerald suggested.

WHAT'S THE TIME?

Isaac Newton wrote that 'Time exists in and of itself and flows equably without reference to anything external.' In Newton's view, time always ticked by at the same pace wherever you measured it. If our timepieces are both accurate, then ten seconds for me will be ten seconds for you too.

James Clerk Maxwell (see page 96) had determined that the speed of an electromagnetic wave is not measured relative to anything else but is a constant defined by the properties of the vacuum of space through which the waves move. Most things may be relative, but the speed of light was absolute, determined by the very nature of the universe. Because Maxwell's equations hold true in any inertial frame, two observers moving relative to each other, each measuring the speed of a beam of light relative to themselves, will both get the same answer – even if one is moving in the same direction as the beam of light and one away from it. The special theory of relativity derives from that one simple fact, that the speed of light is a constant for all observers.

Imagine sending a laser signal to a spacecraft that's heading away from you at half the speed of light. Common sense tells you that the laser light should reach the spacecraft at half-light speed because it has to catch up with the spacecraft, but common sense is wrong. The beam will still arrive at the spacecraft at approximately 300,000 km/s. Newtonian physics tells us that velocity equals distance travelled divided by the time taken to cover that distance – $v = d/t$. So, if the speed of light, v, always remains the same, whatever the other two values, then it follows that d and t, space and time, must change. For you and the pilot in the spacecraft to agree on the speed the beam reaches the spacecraft you have to agree on the time it takes to get there. Since the speed of light remains constant it follows that the spaceship's clocks have to be running slower.

Time, Einstein declared, passes differently in all moving frames of reference, which means that observers in relative motion will have clocks that run at different rates. Time is relative.

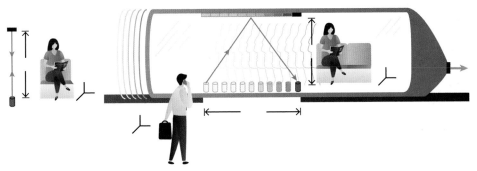

For the observer by the trackside time passes faster than for the observer on the train as the reflected light beam travels further but at the same velocity.

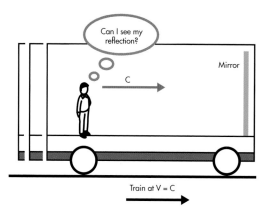

Einstein asked himself if he would see his reflection in a mirror while moving at lightspeed. He concluded that he would.

According to special relativity, the faster you travel through space, the slower you travel through time. Approaching the speed of light the intervals between events lengthen, so time seems to slow down. This phenomenon is called time dilation. If an object could achieve the speed of light, then time would appear to stop entirely. As Werner Heisenberg described it: 'this was a change in the very foundation of physics'. Scientists at CERN's Large Hadron Collider, where particles are smashed together at near lightspeed velocities, have to take the effects of time dilation into account when interpreting the results of their experiments.

GO FASTER, GET SHORTER

Einstein made no mention of Michelson and Morley's experiment in the Special Theory and actually claimed later that he hadn't heard of them when he was formulating it. In 1931, however he publicly addressed Michelson declaring that: 'It was you who led the physicists into new paths, and through your marvellous experimental work paved the way for the development of the theory of relativity.'

Einstein asked himself a question: If I held a mirror while travelling at close to the speed of light, would I see my reflection? How would light reach the mirror if the mirror was travelling at light speed? It was *gedankenexperiment*, or thought experiments, like this that allowed Einstein to lay the foundations of relativity. If the speed of light is a constant, then, no matter how fast he is going, the light going from Einstein to the mirror and back again will always be travelling at 300,000 km/s, because the speed of light does not change. In order to make the calculation work, not only does time have to be slowed down, but also the distance travelled by the light beam has to decrease.

Imagine a spacecraft with mirrors mounted at each end between which a pulse of light bounces. What happens to the pulse as the spacecraft approaches the speed of light?

For a 150-m long ship at rest, the return journey for the light beam will take roughly a millionth of a second. However, at 99.5 per cent of the speed of light, time is slowed by around a factor of ten, which would mean the round-trip journey time, as measured by an observer, is now just a hundred thousandth of a second. However, the pulse from the back to the front has further to go, because the mirror is retreating from it at close to the speed of

Inside the Large Hadron Collider; scientists have to take relativistic effects into account in assessing the results of experiments.

light. The return trip from front to back is shorter because the rear mirror is rushing towards the light beam. But no matter whether the mirror is retreating or advancing, the light beam will always reach it at the same speed, roughly 300 million m/s, because the speed of light doesn't change. In order to keep things in balance, not only does time have to slow down, but also the distance travelled by the light beam has to decrease.

At roughly 99.5 per cent of light speed, the distance is reduced by a factor of ten – the same proportion as the time dilation effect. This shrinkage only takes place in the direction of motion and will only be apparent to an observer who is at rest with respect to the moving object. The crew of a spaceship travelling at close to light speed won't perceive any change in their ship's length, or in themselves, rather they would see the observer appear to contract as they streaked by.

A consequence of the contraction of space is to shorten the time it would take to travel to the stars. Imagine a cosmic railway network with tracks stretching from star to star. The faster a spacecraft travels, the shorter the track appears to become, and therefore the shorter the distance to be covered to reach its destination. At 99.5 per cent of light speed the journey to the nearest star would take around five months in ship time. However, for an observer back on earth the trip would appear to take over four years!

ABSOLUTE SPACETIME

Einstein argued that absolute time and absolute space should be replaced by absolute spacetime. The mathematics of relativity demonstrates that space and time are inextricably linked, and both are altered when we approach near-light speeds. Only by considering space and time together can we give an accurate description of what is observed at light speed.

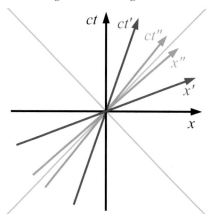

In simple terms a Minkowski diagram maps out all the possible points in spacetime an object can reach in its lifetime.

Motion, distance and the duration of time, all take place against the unchanging backdrop of spacetime, the geometry of which is rigidly dictated by the speed of light. Absolute spacetime is as crucial to the understanding of special relativity as the absolute time and absolute space it replaced were to Newtonian physics.

In 1907, mathematician Hermann Minkowski (1864–1909) developed a way of visualizing how objects moved through space and time. These are called Minkowski spacetime diagrams and they provide a graphical way of visualizing some of the odd effects of relativity. A Minkowski diagram employs a coordinate

Hermann Minkowski, at one time Einstein's maths teacher, said he was 'a lazy dog' as a student!

system, with time shown vertically along the y axis, and space dimensions represented along the x and z axes. An object is not represented as a single point but as a line describing all the spacetime points at which it exists. This is the object's worldline.

Spacetime diagrams can be used to explain some of the puzzling effects of special relativity, such as time dilation and length contraction. In the old Newtonian physics, travel through time and travel through space were held to be two quite separate things. But, according to Einstein, this just isn't the case. According to special relativity, the combined speed of an object's motion through time and through space is precisely equal to the speed of light. This is an upper speed limit that can't be broken. For an object in motion, time *must* slow down otherwise the total combined speed through spacetime would exceed the speed of light. At the speed of light all the spacetime movement has become movement through space with nothing left over for movement through time.

Think of an aircraft flying due south. If it changes course, but not speed, so it is now flying southwest, it is still going in a southerly direction but not as quickly as before because part of its velocity is now taking it in a westerly direction as well. For 'south', substitute 'time' and for 'west' substitute 'space' and the situation is analogous. If an object is stationary – that is, not moving through space – then all of its spacetime movement is through time. If it moves through space, its movement through time is slowed as some of its spacetime motion is used for its journey through space.

These relativistic effects apply for any movement through space, even at slow speeds. Experiments carried out using atomic clocks have demonstrated that clocks flown on an aircraft ran a few hundred billionths of a second slower than similar clocks that remained on the ground. It was a small difference, but it matched exactly with the predictions of special relativity.

$E=mc^2$

One of the most famous equations in all physics is derived from Special Relativity. Einstein published it as a sort of postscript to the special theory in a short paper, just three pages long, entitled 'Does the Inertia of a Body Depend Upon its Energy Content?' in September 1905. Known sometimes as the law of mass-energy equivalence, $E=mc^2$ effectively says that energy and mass are two aspects of the same thing. If an object gains or loses energy, it loses or gains an equivalent amount of mass in accordance with the formula. For example, the faster an object travels, the greater is its kinetic energy, and the greater also is its mass. The speed of light is a big number – squared it is a very big number indeed. This means that when even a tiny amount of matter is converted into its energy equivalent the yield is colossal, but it also means that there has to be an immense input of energy to see an appreciable increase in mass.

Chapter 10

BENDING SPACETIME

Bending Spacetime

THEORIES OF SPACETIME

Timeline	
1907	Albert Einstein hypothesizes that gravity and acceleration are equivalent.
1915	Einstein sets out the theory of general relativity.
1916	Einstein is able to use relativity to explain anomalies in Mercury's orbit.
29 MAY 1919	Sir Arthur Eddington's observation of a solar eclipse confirms Einstein's prediction of the bending of light by gravity.
1962	In an experiment involving two atomic clocks, Einstein's prediction of gravitational time dilation is proved to be correct.
1974	Observation of a binary pulsar by the Arecibo Radio Observatory provides the first evidence for the existence of gravitational waves.
2015	LIGO detects gravitational waves for the first time.

In order to make the calculations easier, Einstein had limited the special theory of relativity to objects in uniform motion. For the sake of simplicity, he had ignored the effects of acceleration and gravity. Bringing gravity into the general theory of relativity would take seven years of intense work in what physicist Dennis Overbye described as 'arguably the most prodigious effort of sustained brilliance on the part of one man in the history of physics'. What emerged would change our whole conception of the way the universe worked.

FEELING GRAVITY'S PULL

How did gravity make its influence felt? Obviously, it was a force that acted at a distance and didn't require any physical contact to work. Also, unlike any other force, it was impossible to shield yourself from its effects. Newton's universal law of gravity had been backed up by observation and experiment time and time again. According to Newton, gravity made its effects felt instantaneously. If the sun were to suddenly disappear, the earth would shoot out of its orbit in the same second. There would be no waiting eight minutes for the last rays of light to arrive.

As Einstein and Galileo had shown, if you are in uniform motion then it's impossible to demonstrate that you are actually moving. All observers moving uniformly relative to each other are entitled to say that they are stationary and that it's everyone else who is moving.

Accelerated motion is quite different. If we change speed or direction, we *feel* it. Accelerating results in inertial forces – the forces that resist a change in speed or direction. These are the forces that tilt you to one side when your car takes a bend in the road.

Galileo had demonstrated that a small stone will fall to the ground in the same time it takes a large stone to do so. This is because the two stones accelerate towards the ground at the same rate, something Newton explained with his second law of motion – force equals mass times acceleration. The idea that gravity accelerates all objects at the same rate regardless of what they might be made of is called the 'Universality of Free Fall' or the 'Equivalence Principle'. In Newtonian theory, the inertial mass of a body, its resistance to acceleration, and its gravitational mass, determined by the strength of the gravitational force acting on it, matched exactly. Einstein believed that this couldn't be coincidence.

Towards the end of 1907 Einstein had what he called his 'happiest thought'. He realized that gravity and acceleration were equivalent; without a frame of reference it was impossible to tell one from the other. Lecturing in Kyoto, Japan, in 1922 he said: '... all of a sudden a thought occurred to me: *If a person falls freely he will not feel his own weight.* I was startled. This simple thought made a deep impression on me. It impelled me towards a theory of gravitation.'

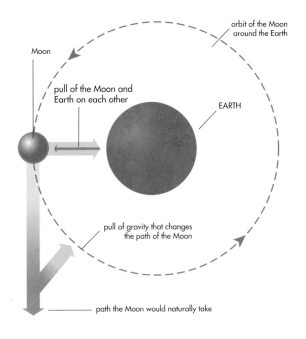

Newtonian gravity.

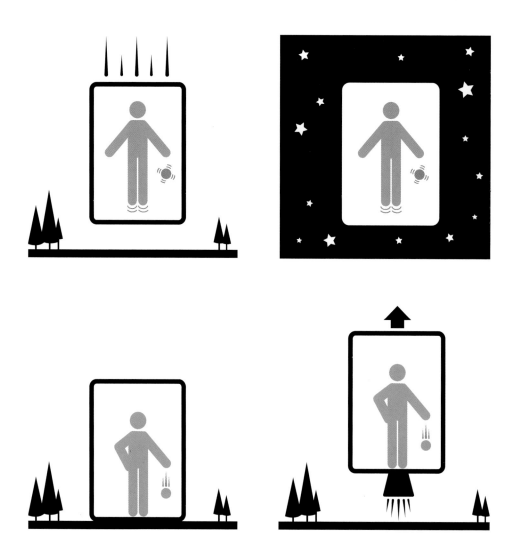

Einstein's equivalence principle states that the effects of gravity and acceleration are indistinguishable.

In another of his thought experiments, Einstein developed the idea along these lines: Imagine waking up in a box. Unknown to you, the box is in deep space and under uniform acceleration. If you drop objects in the box, their inertia causes them to fall to the 'bottom' of the box, which is, in the opposite direction to that in which the box is moving. All objects will fall in exactly the same way, no matter what their mass or composition, in accordance with the laws of Galileo and Newton. From your observations it would be reasonable to conclude that there was a gravitational field at work inside the box.

Einstein's assertion was that the acceleration wasn't just producing an effect that was *similar* to a gravitational field, it actually *was* a gravitational field. He formulated a principle of equivalence that stated that the effects of uniform acceleration were indistinguishable from the effects of gravity. Acceleration creates a gravitational field. According to Einstein's equivalence principle, whether the person in the box was accelerating or not depended on your point of view. An observer outside the box would see it accelerating uniformly through space, the person inside the box would consider themself to be in a gravitational field. Each point of view is equally valid. This was what made the inertial mass and the gravitational mass the same. Tests of the general theory's equality between gravitational and inertial mass have been found to be accurate to within one part in 10 trillion, which is as accurate as we can possibly be with current equipment.

PHOTON SHIFTS

Einstein's equivalence principle predicts that gravity will have an effect on the wavelength of electromagnetic radiation. In accordance with Einstein's $E=mc^2$, and Planck's $E=\hbar f$ law (see page 141) relating the energy of light to its frequency, it becomes apparent that a photon moving out of a gravitational field must lose energy. Since photons always travel at the speed of light, this energy loss takes the form of a lowering in frequency, rather than as a reduction in velocity. A beam of light projected from the surface of the earth will be lower in frequency when it reaches an observer in orbit. This lowering of the frequency of the photon corresponds to a 'redshift' to the lower-frequency, longer-wavelength end of the spectrum.

Another prediction of the equivalence principle is that the path of a light beam will be bent by gravity. Imagine a photon crossing inside the box as it accelerates through space. As the photon is crossing the box, the floor is accelerating upwards, which means that the photon appears to fall downwards. Because a gravitational field is equivalent to acceleration, the same must also hold true there.

A consequence of this frequency shift is the slowing of time. A lower frequency means that the length of time between one wave crest and the next increases. It would appear to the orbiting observer that things down below were taking a little longer to happen. This effect, called gravitational time dilation, means that observers at different distances from a large object (which produces a gravitational field) will obtain different measurements for the time elapsed between two events. This is a direct consequence of the fact that the observer outside the accelerating box, that is, one outside the gravitational field, sees the photon follow a straight path, but the observer in the box sees it follow a longer, curved path. Because the speed of light cannot change, a clock inside the box has to run slower than one outside it to allow both journeys to be made in the same time.

This prediction of general relativity was tested in 1962 when two extremely accurate atomic clocks were placed on a tower, one at the top and one at the bottom. The clock at the bottom, the one deepest in the earth's gravity well, ran slower than the one at the top. The discrepancy was exactly in line with Einstein's prediction.

EINSTEIN'S GRAVITY

According to Newton, gravity is a force acting between objects that pulls them towards each other. Einstein's theory of gravity approached things in a different way. Rather than exerting a force, he suggested that a mass causes a distortion of spacetime. Empty spacetime, the spacetime of special relativity, is flat. But where matter is present, spacetime is curved. In the same way that there are no straight lines on the surface of a sphere, there are no straight lines in curved spacetime. The closest we can get to the straight line in curved spacetime is a geodesic, a curve that is as straight as possible. The meteoroid being pulled towards a planet has not been deflected from its straight-line course through space, rather the presence of the planet has distorted spacetime, changing the form a straight line can take. It has redefined the geometry of spacetime. The planet makes a dent in spacetime, curving it around itself.

In Einstein's universe, gravity is the result of curved spacetime. Objects still follow the straightest possible paths through spacetime, but because spacetime is now curved, they accelerate as if they were under the influence of a gravitational force. Matter distorts the geometry of spacetime and this distorted geometry dictates how matter moves through it. As the matter moves and the sources of gravity change positions, so the swirling curves of spacetime ebb and flow also. As physicist John Archibald Wheeler succinctly summarized it: 'Spacetime tells matter how to move; matter tells spacetime how to curve.'

PERIHELION PUZZLE

A long-standing puzzle for astronomers was the fact that the orbit of Mercury, the planet closest to the sun, did not quite fit in with Newton's equations. As the planets orbit the sun, they follow an elliptical path, as determined by Kepler in 1609 and explained by Newton some 50 years later. The elliptical path means that there is a point on the planet's orbit at which it makes its closest approach to the sun (astronomers call this the perihelion). This point doesn't always occur at the same place on each orbit. Because of the pull of the planets on each other, an effect predicted by Newton, the perihelion slowly moves around as the planet orbits the sun. This rotation of the orbit is called a precession.

The problem was that Newtonian physics could explain the precession of all the planets bar Mercury. Mercury's rate of precession was just a little bit more than Newton predicted. It was a small difference but not one that could be ignored.

According to Einstein, gravity is a result of mass curving spacetime around itself.

Astronomers hunted for a way to explain Mercury's odd behaviour. Perhaps there was a swarm of asteroids between Mercury and the sun, or perhaps even an undiscovered planet, that was tugging on Mercury as it orbited. There were many ideas, but none seemed to answer all the questions. What they all had in common though was that they accepted Newton's law of gravitation as accurate. On Christmas Eve 1911, Einstein wrote: 'I am [again] busy with considerations on relativity theory in connection with the law of gravitation... I hope to clear up the so-far unexplained changes of the perihelion length of Mercury... [but] so far it does not seem to work.'

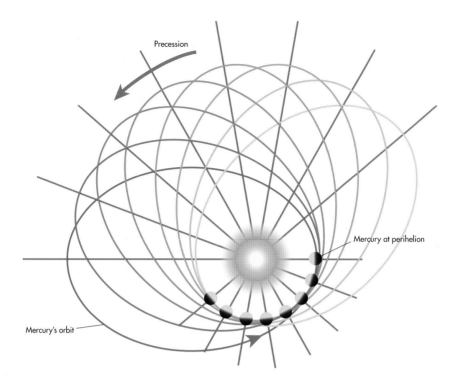

Einstein's theory of gravity accounted for anomalies in the orbit of Mercury that Newtonian gravity failed to explain.

By 1916, Einstein was ready to step in with the equations from his newly-forged general theory of relativity. He was able to show that his concept of how gravity worked exactly predicted Mercury's orbital movements. The reason for the discrepancy was the warping of spacetime so close to the huge mass of the sun. Einstein was jubilant at this evidence that his theory was right and his calculations agreed with the astronomers' observations.

NEWTON ECLIPSED

When Einstein first set out his equivalence principle in 1907 and deduced that it would result in the bending of light, he thought that the effect would be far too small ever to be measured. At this time, Einstein had not yet arrived at the insight that spacetime was curved and that this would have an effect on the bending of the light beam, so his first predictions for the bending of light were in accord with what Newton himself would have predicted from his law of gravitation and his belief that light took the form of a stream of particles.

It wasn't until 1915 that Einstein realized that, according to his general relativity theory, light rays would be bent by twice the value of his initial 1907 calculation. There was now

Asked if it was true only three people understood general relativity, Arthur Eddington allegedly quipped, 'Who is the third?'

a clear difference between Einstein's general relativity predictions and those of Newtonian physics. Einstein was anxious to have his theory proved right.

Astronomer Sir Arthur Eddington obtained a copy of Einstein's theory in 1916 and became an enthusiastic champion of relativity. Together with the Astronomer Royal, Sir Frank Dyson, he came up with a plan to test Einstein's theory. During the day the faint light from the stars is normally drowned out by the far greater radiance of the sun, but during a total solar eclipse, when the moon temporarily blocks the light from the sun, the stars become briefly visible in the daytime sky. Einstein predicted that light that passed close to the sun on its way to earth would have its path deflected by the warped spacetime around the sun. This would result in a change in the apparent position of a star from its actual position, which was known from observations of its position at night. The angle of deflection was very small indeed, roughly equivalent to the width of a coin seen from a distance of 3 kilometres.

An image of the 1919 eclipse that confirmed Einstein's prediction of light bending.

Determined to make the necessary measurements, Eddington led an expedition to the island of Principe, off the coast of West Africa, to observe the total eclipse of 29 May 1919. A back up expedition was also sent to Brazil. Eddington reported that he didn't actually see the eclipse beyond a couple of glances as he was too busy changing the photographic plates in his camera. One of the Brazil expedition photographs appeared to agree with Einstein, another with Newton. Eddington's photographs showed fewer stars, but seemed to back Einstein. Eddington decided that the Newton-supporting photograph from Brazil was due to faulty equipment

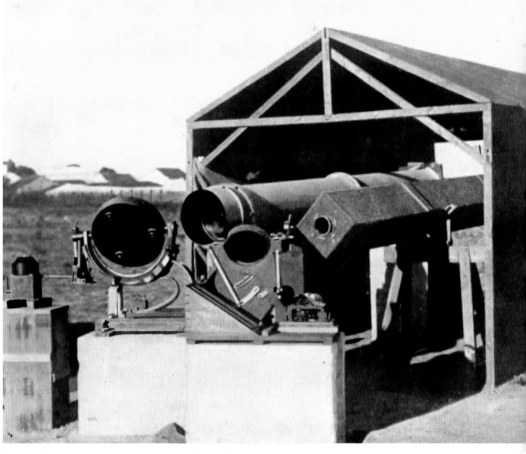

Some of the equipment employed by the Brazilian eclipse expedition.

and declared Einstein vindicated. Famously, soon after he heard the news, Einstein was asked what he would have done if the observations had shown his theory to be wrong. Einstein replied: 'Then I would have been sorry for the dear Lord; the theory is correct.'

GRAVITATIONAL WAVES

The predictions of the general theory of relativity are the same as those of Newton's theory of gravitation so long as the gravitational fields involved are weak; in other words, so long as the velocities of all objects interacting with each other gravitationally are small compared with the speed of light. A gravitational field is considered strong if the escape velocity required to break free of it approaches the speed of light. All gravitational fields encountered in the solar system, even the one in the vicinity of the sun, are weak by this definition. At low speeds and

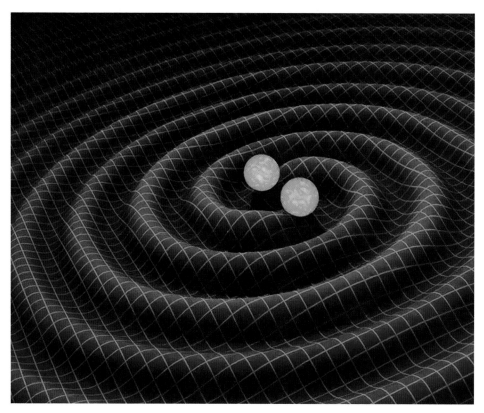

Massive objects orbiting each other, such as a pair of pulsars, send gravitational waves rippling through spacetime.

in weak gravitational fields, general and special relativity's predictions agree with everyday experience and Newtonian physics.

One of the predictions of general relativity was a phenomenon called 'gravitational waves'. Gravitational waves are like ripples in spacetime caused by particularly energetic disturbances. Einstein's equations showed that cataclysmic events, such as a collision between two black holes, would be like a large rock being dropped into the pond of spacetime, sending out waves of distorted space across the universe at the speed of light.

Although gravitational waves were predicted to exist in 1916, there was no actual proof of their existence until 1974, when astronomers at the Arecibo Radio Observatory in Puerto Rico discovered a binary pulsar – two extremely dense and heavy stars in orbit around each other. The astronomers began making careful observations of the system and after eight years of meticulous data-gathering established that the pulsars were getting closer to each other at exactly the rate general relativity predicted that they would. After over four decades of close monitoring, the observed changes in the orbits of the pulsars is in such close agreement with general relativity that the researchers have no doubt that the system is emitting gravitational waves.

The 305-metre diameter Arecibo radio telescope explores the universe from the edge of space to the furthest reaches of the cosmos.

Until 14 September 2015, all the confirmations of the existence of gravitational waves had been indirect or determined mathematically, and not through actual physical proof. On that day, the Laser Interferometer Gravitational-Wave Observatory (LIGO) in the United States detected gravitational waves for the first time. The waves it detected were generated by two colliding black holes nearly 1.3 billion light years away. Although this was an extremely violent event, by the time the waves reached the earth the degree of spacetime wobbling they generated was much smaller than the nucleus of an atom. Because the fluctuations are so small, LIGO has to be extremely sensitive and is an absolute triumph of engineering skill and ingenuity. Two L-shaped detectors built 3,000 km apart (in the US states of Washington and Louisiana) and housed inside 4-km-long vacuum chambers work in unison to measure a motion 10,000 times smaller than an atomic nucleus. It is a degree of accuracy equivalent to measuring the distance to the nearest star to within less than the width of a human hair.

The gravitational wave detector is one of the most sensitive instruments ever built.

Chapter 11

THE QUANTUM REALM

The Quantum Realm

TIMELINE OF QUANTUM PHYSICS

Timeline	
1801	Thomas Young performs the double-slit experiment for the first time.
1887	Heinrich Hertz discovers the photoelectric effect.
1905	Albert Einstein establishes that light consists of discrete particles, or energy quanta, later called photons.
1922	Arthur Compton discovers that X-rays also act like particles when he uses them to examine the distribution of electrons within an atom.
1924	Louis de Broglie theorizes that all matter and energy has both waves and particle characteristics.
1925	Max Born imagines electrons as waves of probability around the atom.
1925–7	Niels Bohr and Werner Heisenberg develop the Copenhagen interpretation of quantum mechanics.
1926	Erwin Schrödinger develops an equation to determine how wave functions are shaped.
1927	Heisenberg sets out his uncertainty principle, putting a limit on what we can know.
1935	Albert Einstein, Boris Podolsky and Nathan Rosen attempt to demonstrate the flaws of quantum mechanics through the EPR paradox.
1947–9	Richard Feynman, Julian Schwinger and Tomonaga Shin'ichirō develop the theory of quantum electrodynamics.
1980–2	Alain Aspect conducts a series of experiments proving the truth of the phenomenon of quantum entanglement.

On 19 October 1900, Max Planck made a presentation to the Deutsche Physikalische Gesellschaft in Berlin in which he submitted a radical solution to a problem that had been vexing physicists up until then. Although it would be a few years before the full implications of what he had to say became clear, his pronouncements ushered in a new age for physics – the era of the quantum.

Theory predicted that the electromagnetic radiation emitted from a blackbody (see page 112), a perfect absorber and reflector of radiation, should become infinite for shorter and shorter wavelengths. Everyday observation showed that this prediction, which was dubbed the ultraviolet catastrophe, was obviously wrong – bakers weren't being lethally irradiated every time they opened their ovens – but no one could come up with a theory that explained why.

Planck made the revolutionary suggestion that the energy emissions of the blackbody, rather than being a continuously variable quantity like a wave, in fact came in discrete packets, which he called quanta (singular quantum) from a Latin word meaning 'how much'. The size of these quanta was proportional to the frequency of vibration and energy could only be emitted or absorbed in whole quanta. This explained why a blackbody did not give off energy equally across the electromagnetic spectrum. Although there were an infinite number of higher frequencies in theory, it took increasingly large amounts of energy to release quanta at that level. It was twice as easy to release a quantum of red light, for instance, than it was to release a quantum of violet light with twice the frequency.

Planck's constant

Planck suggested that the energy of a quantum was related to its frequency using the simple formula $E = \hbar f$ where E equals energy, f equals frequency and $\hbar$ equals a value known as Planck's constant. The energy of a quantum can be calculated by multiplying its frequency by Planck's constant, which is $6.62607015 \times 10^{34}$ J/s.

There was no doubt that Planck's solution worked – the results of experiments were in accord with the predictions made by his theory. Nonetheless, Planck wasn't entirely happy with his explanation; it flew in the face of all he had previously learned about physics and he resisted for years the idea that his quanta had any basis in reality. Rather, he saw them as more of a mathematical 'fix' for a difficult problem. By his own admission he had introduced them as 'an act of desperation' and looked for ways to disprove his own theory. It would be some years before the unforeseen consequences of Planck's desperate act made themselves felt in the world of physics. When Albert Einstein heard about Planck's theory he commented: 'It was as if the ground had been pulled out from under us, with no firm foundation to be seen anywhere.'

THE PHOTOELECTRIC EFFECT

In 1904, Einstein wrote to a friend that he had discovered 'in a most simple way the relation between the size of elementary quanta ... and the wavelengths of radiation'. This relationship was the answer to a curious aspect of radiation that had previously defied explanation. In 1887, German physicist Heinrich Hertz (1857–94) had discovered that certain types of metal would emit electrons when a beam of light was directed at them. This was the 'photoelectric effect'. At first it was thought that the effect could be explained in terms of electromagnetism. The electric field part of the electromagnetic wave, it was assumed, gave the electrons the energy they needed to break free from the metal. But it soon became apparent that this couldn't be the whole story.

Heinrich Hertz, discoverer of the photoelectric effect.

If the theory was correct, then the brighter the light the higher energy the emitted electrons should be, but experiment showed that the energy of the electrons released depended on the *frequency* of the light – not its *intensity*. No matter how bright the light, the electrons that emerged still had the same energy. It was only by shifting the light up in frequency, from red to violet and then ultraviolet, that higher energy electrons were emitted. If the light was of a low enough frequency no electrons would be emitted at all, even if the light was blindingly bright. It was as if fast-moving ripples could readily move the sand on a beach but a slow-moving wave, no matter how big, left them untouched. In addition, if the electrons were going to jump at all, they jumped right away – there was no build up of energy involved. The wave theory of light could make no sense of these findings.

In March 1905, Einstein published a paper in the journal *Annals of Physics* that would eventually win him the Nobel Prize in 1921. What he did was to look at the differences between particle theories and wave theories, comparing the formulae that described the way the particles in a gas behaved as it changed volume with those describing the way that waves of radiation spread through space. He found that both obeyed the same rules and the mathematics underpinning both phenomena was the same. Einstein wrote that: 'When a light is propagated from a point, the energy... consists of a finite number of energy quanta which are localized at points in space and which can be produced and absorbed only as complete units.' Einstein's biographer Walter Isaacson described this as 'perhaps the most revolutionary sentence that Einstein ever wrote'.

Einstein used these insights to calculate the energy of a light quantum of a particular

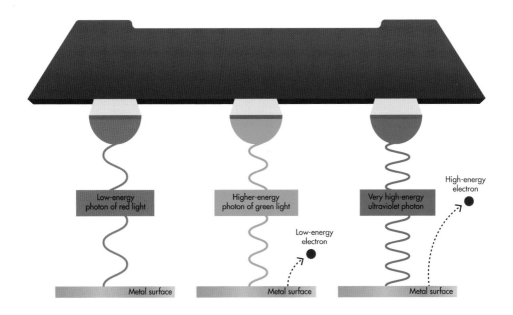

The photoelectric effect depends on the frequency of light, not its intensity.

frequency. He found that his results agreed with Planck's and went on to show how the existence of light quanta could be used to explain the photoelectric effect. As Planck had established, the energy of a quantum was determined by its frequency. A single quantum transferred its energy to an electron – the higher the energy of the quantum, the more likely it was to cause the electron to be emitted. High-energy blue quanta had the heft to punch out electrons, low-energy red quanta simply didn't have the muscle.

Whereas previously Planck had been considering the quantum as little more than a mathematical contrivance, Einstein was now suggesting that it was an actual physical reality. It was a suggestion that didn't go down well with other physicists who were reluctant to give up the idea that light was a wave, and not a stream of particles. Even Planck suggested that 'he may have gone overboard in his speculations'. A sceptical Robert Millikan (1868–1953) performed experiments in 1915 on the photoelectric effect that were aimed at disproving Einstein's assertion, and ended up producing results that were entirely in line with Einstein's predictions, although this didn't prevent Millikan from continuing to refer to Einstein's 'reckless hypothesis'.

THE NATURE OF LIGHT

Scientists were faced with having to rethink their ideas on the nature of light. Einstein had established that light acted as if it were a stream of particles, but centuries of experiment

Arthur Compton, who confirmed that X-rays had particle-like properties.

and observation had also shown beyond doubt that it acted like a wave in familiar and well understood phenomena such as diffraction and interference. So now the question was this: what *was* light? Was it a wave or was it a particle? Einstein himself struggled without success to solve the dual nature paradox of light. Towards the end of his life, in 1951, Einstein wrote in a letter to his friend Michele Besso that 'fifty years of conscious brooding have brought me

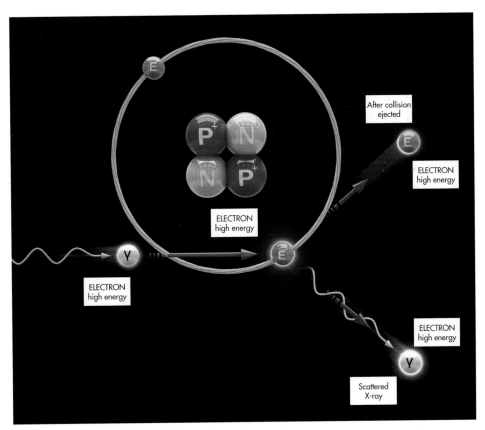

The Compton effect demonstrated that collisions took place between X-rays and electrons.

no closer to the answer to the question, "What are light quanta?" Of course, today every rascal thinks he knows the answer, but he is deluding himself.'

In 1922 American physicist Arthur Compton (1892–1962) carried out an experiment in which he used X-rays to investigate how electrons were distributed in atoms. He discovered that the X-rays had lower frequencies and longer wavelengths after interacting with the electrons, which meant that they had lost energy. The small change in the frequency of the X-rays became known as the Compton effect. Compton established that the X-rays and the electrons were acting like particles when they collided and provided proof that Einstein was right – light could indeed behave like a particle. In a paper published in the journal *Physical Review*, Compton wrote 'this remarkable agreement between our formulas and the experiments can leave but little doubt that the scattering of X-rays is a quantum phenomenon.' Although he wasn't the first to use it, Compton established the name *photon* for a quantum of light.

French physicist Louis de Broglie (1892–1987) put forward a theory in his doctoral thesis, *Recherches sur la Théorie des Quanta* (Researches on the quantum theory) in 1924, in which he proposed that not just light, but all matter and energy has the characteristics of both particles and waves. Believing intuitively in the symmetry of nature and Einstein's quantum theory of light, de Broglie asked, if a wave can behave like a particle then why can't a particle, such as an electron, also behave like a wave? De Broglie reasoned that as Einstein's famous $E=mc^2$ relates mass to energy and Einstein and Planck had related energy to the frequency of waves, then combining the two suggested that mass should have a wavelike form as well.

De Broglie came up with the concept of a matter wave, suggesting that any moving object had an associated wave. The kinetic energy of the particle is proportional to its frequency and the speed of the particle is inversely proportional to its wavelength –

Louis de Broglie proposed that all matter and energy can behave both as wave and particle.

faster particles having shorter wavelengths. Einstein supported de Broglie's idea as it seemed a natural continuation of his own theories, commenting that 'de Broglie's hypothesis is the first feeble ray of light on the worst of our physical enigmas.' De Broglie's thesis was experimentally verified in 1927 when George Thomson in the UK and Clinton Davisson in the United States both demonstrated that a narrow electron beam directed through a thin crystal of nickel formed a diffraction pattern as it passed through the crystal lattice.

At the beginning of the 19th century Thomas Young (see page 57) had convincingly demonstrated that light was a wave by showing how it formed interference patterns when it passed through twin slits. In the early 1960s Richard Feynman (1918–88) described an experiment in which he imagined what would happen if just one photon or electron at a time was directed towards a detector

Clinton Davisson (left) and Lester Germer, who showed that a stream of electrons forms a diffraction pattern.

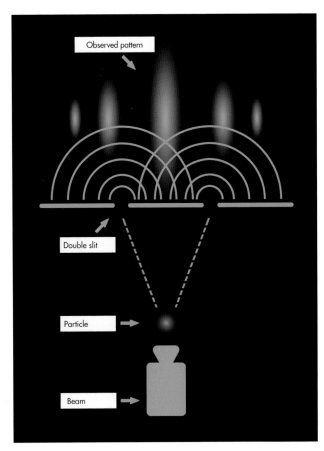

Observed pattern

Double slit

Particle

Beam

The double-slit experiment, described by Richard Feynman as having within it 'the heart of quantum mechanics'.

through twin slits that could be opened or closed. Common sense would appear to dictate that the photons would travel as particles, arrive as particles and be detected on the screen as individual dots. Rather than interference patterns, there should be two bright areas when both slits were open or just one if one slit was closed. However, what happens is that the pattern on the screen builds up, particle by particle, into interference patterns when both slits are open – but not if one slit is closed. Even if subsequent photons are fired off *after* the earlier ones have hit the screen they still somehow 'know' where to go to build up the interference pattern. It is as if each particle travels as a wave, passing through both slits simultaneously, and interferes with itself!

How can that be? How does a single particle travelling through the left-hand slit know whether the right-hand slit is open or closed? Feynman advised against even attempting to answer these questions, declaring it *'absolutely* impossible, to explain in any classical way'. In 1964 he wrote: 'Do not keep saying to yourself, if you can possibly avoid it "But how can it be like that?" because you will go down the drain into a blind alley from which nobody has yet escaped. Nobody knows how it can be like that.'

What was apparent was that, bizarre as it might seem, both the wave theory and the particle theory of light were correct. Whether light acts as a wave or as a particle seems to depend on how it is being measured. We have no single model that can describe light in all its aspects. It's easy enough to say that light has 'wave–particle duality' and leave it at that, but what that actually means is something that no one can answer satisfactorily.

Richard Feynman, one of physics' most original thinkers.

PROBABILITY WAVES

The experiments carried out in the 1920s had established fairly conclusively that electrons and other particles could act like waves, but what was the nature of those waves?

In 1802, astronomer William Hyde Wollaston had noticed that the spectrum of sunlight was overlaid by a number of fine black lines. The lines were examined in detail by German lens-maker Joseph von Fraunhofer, whose name they now bear, and in the 1850s Gustav Kirchhoff and Robert Bunsen established that each element produces its own unique set of lines, but what caused them was a mystery.

Albert Einstein's explanation of the photoelectric effect imagined light behaving like a stream of quanta, which, if powerful enough could knock electrons out of atoms. In 1913, Danish physicist Niels Bohr proposed a model of the atom that accounted for both Einstein's quanta and the spectra of elements.

In Bohr's atom electrons travelled around the core nucleus in fixed, or quantized orbits. Photons, light quanta, striking the atom could be absorbed by electrons, which then moved to higher orbits, further from the nucleus of the atom. A sufficiently energetic photon could eject an electron from its orbit altogether. Photons were released by the transition of electrons downward in their orbits. These steps up and down energy levels are what we now call quantum leaps.

Atoms give off light only at very specific wavelengths, which means that each element produces a characteristic set of spectral lines. Bohr proposed that these lines were related to the energies of the electron orbits and were produced by an electron absorbing or emitting a photon at the frequency corresponding to the spectral line.

In June 1925, German physicist Werner Heisenberg made a breakthrough. Rather than consider the electrons of an atom to be in fixed orbits as Bohr had done, he conceived of them as representing the harmonics of a series of standing waves. The equations he formulated linked these waves to the quantum leaps of the electron from one orbit to another. Heisenberg's fellow physicist Max Born (1882–1970) saw that Heisenberg's idea could be used to link the energies of electrons to the lines that had been observed in the visible light spectrum. According to Born, the wave nature of matter meant thinking in terms of

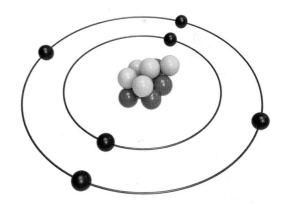

Niels Bohr's 1913 model of the atom imagined electrons in fixed orbits around the nucleus.

Werner Heisenberg saw electrons as being like waves around the nucleus rather than orbiting it.

probabilities rather than certainties. He described the electron wave as being like a graph mapping out the probability of finding the electron in a particular place. This is one of quantum physics' stranger ideas. How can a particle be perhaps here, but also perhaps there?

Austrian physicist Erwin Schrödinger worked out an equation in 1926 that determined how probability waves, or wavefunctions as they came to be known, are shaped and how they evolve. The wavefunction holds all of the information necessary to describe a quantum system. The Schrödinger equation is as important to quantum mechanics, which deals with the mathematical description of the interactions of subatomic particles, as Newton's laws of motion were for forces and motion on a large scale. Schrödinger was describing the quantum world in purely mathematical terms, with outcomes that could only be seen in terms of probabilities and not certainties. This was a world beyond anything that could be visualized in terms of 'real world' analogues.

One of quantum mechanics' biggest shortcomings was its failure to take Einstein's relativity theories into account. One of the first to try reconciling the two was British physicist Paul Dirac (1902–84). In 1928, he succeeded in marrying Schrödinger's equation to Einstein's famous E=mc² equation and produced an equation that was consistent with both special relativity and quantum

Erwin Schrödinger determined an equation to describe the probability waves of the quantum world.

mechanics in its description of electrons and other particles. The Dirac equation, which viewed electrons as excitations of an electron field, in the same way that photons could be seen as excitations on the electromagnetic field, became one of the foundations of quantum field theory. Physicist Freeman Dyson thought of it as a significant step forward in our understanding of reality, describing it as bringing 'a miraculous order into the previously mysterious processes of atomic physics.'

Antiparticles

An unforeseen consequence of Dirac's equation was that it had a solution that suggested the existence of a particle that was the electron's exact opposite. In 1932, this previously unknown particle, the positron, was discovered by Carl Anderson (1905–91). Today, all particles are believed to have their equivalent antiparticles.

AN UNCERTAIN UNIVERSE

Classical physics generally accepted that the accuracy of any measurement was only limited by the precision of the instruments used to make it. In theory, at least, our knowledge of the universe could be as fine-tuned as our technical skills allowed it to be. In 1927, Werner Heisenberg set out to prove that this just wasn't the case.

Heisenberg asked himself what it actually meant to define the position of a particle. We can only know where something is by interacting with it. The position of an electron, for instance, is determined by bouncing a photon off it. The accuracy of the measurement is determined by the wavelength of the photon; the higher the frequency of the photon the more accurately we can determine the position of the electron. However, Planck had shown that the higher the frequency of the photon the more energy it carries and the more likely it is that it will knock the electron off course. We may know where the electron is at the moment of interaction, but we can't know where it is going to be subsequently. If it were possible to measure the electron's momentum with absolute precision its location would become completely uncertain.

Heisenberg showed that the uncertainty in the momentum multiplied by the uncertainty in the position can never be smaller than Planck's constant, linking the energy of a quantum to its frequency. It is a fundamental property of the universe that puts a limit on what we can know.

THE COPENHAGEN INTERPRETATION

Heisenberg had formulated his uncertainty principle while he was a lecturer at Niels Bohr's Institute for Theoretical Physics in Copenhagen. Bohr and Heisenberg brought together their ideas on quantum physics in what became known as the Copenhagen Interpretation.

One of the central strands of the Copenhagen Interpretation is the principle of complementarity. This views the wave and particle nature of objects as complementary aspects of a single reality. An electron or a photon, for example, can behave sometimes as a wave and sometimes as a particle, but can never be seen to be both at the same time, just as a tossed coin can be either heads or tails, but not both simultaneously. The Copenhagen Interpretation treats the wavefunction as no more than a tool for predicting the results of observations, and cautions physicists against concerning themselves with trying to imagine what 'reality' looks like.

Bohr declared that it was meaningless to ask what an electron *really* is. Experiments designed to measure waves will see waves, while experiments designed to measure the properties of particles will see particles. It is impossible to design an experiment that would allow us to see wave and particle at the same time. The wavefunction is a complete description of a wave/particle. When a measurement of the wave/particle is made, its

Quantum tunnelling

Imagine throwing a ball against a wall and watching astonished as it vanished through to the other side instead of bouncing back. The phenomenon of quantum tunnelling allows electrons and other particles to do something very similar, passing through barriers that seem impassable. This oddity arises from considering electrons, for example, as stretched out waves of probability rather than particles existing at a particular point. The Heisenberg Uncertainty Principle forbids us from knowing how much energy a particle has, or its exact location at any precise moment in time. There is a chance, albeit a very small chance, that the electron's probability wave will extend to the other side of the barrier. And sometimes it does. The effect is seen in transistors in which quantum tunnelling allows electrons to pass across a junction between semiconductors. On a more cosmic scale, quantum tunnelling plays a role in the fusion reactions that power stars. Without quantum tunnelling the sun wouldn't shine.

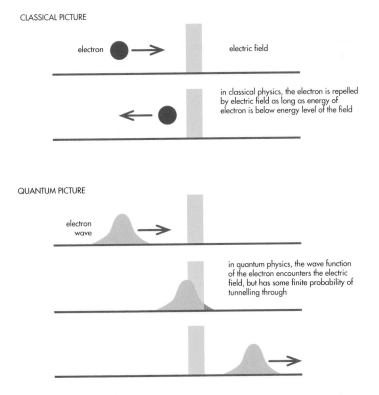

CLASSICAL PICTURE

electron

electric field

in classical physics, the electron is repelled by electric field as long as energy of electron is below energy level of the field

QUANTUM PICTURE

electron wave

in quantum physics, the wave function of the electron encounters the electric field, but has some finite probability of tunnelling through

Uncertainties in the position and energy of a particle/wave allows the phenomenon of quantum tunnelling to occur.

wavefunction collapses. Any information that cannot be obtained from the wavefunction does not exist.

The quantum world as seen through the Copenhagen Interpretation is one of pure statistical probability. The Copenhagen view is that indeterminacy is a fundamental feature of nature and not just something that results from our lack of knowledge. We just have to accept that this is how things are and not try to explain them. It opened up a sharp divide between the deterministic world of classical physics, where every event was assumed to have a cause, and the new quantum world of chance and uncertainty.

The Copenhagen Interpretation was not without its opponents, including Albert Einstein. In an often-quoted letter to Max Born, written in 1926, Einstein wrote: 'Quantum mechanics is certainly imposing. But an inner voice tells me that it is not yet the real thing... I, at any rate, am convinced that *He* does not throw dice.'

Einstein couldn't bring himself to believe that while there appeared to be rules that governed most of what happened in the universe, at the fundamental level of quantum reality things seemed to be left to chance. He held fast to the notion that there existed an objective reality that could be measured, rejecting Heisenberg and Bohr's view that the very act of measurement determined the nature of reality. Einstein believed that Heisenberg's uncertainty principle might very well demonstrate the limits nature places on what we can know, but these limits should not be taken to imply that there wasn't a deeper, more deterministic reality that remained inaccessible to us.

QUANTUM ENTANGLEMENT

In 1935, Einstein, in collaboration with his colleagues Boris Podolsky and Nathan Rosen, explored an aspect of quantum mechanics that disturbed him. In a paper entitled 'Can the Quantum Mechanical Description of Physical Reality Be Regarded as Complete?' they

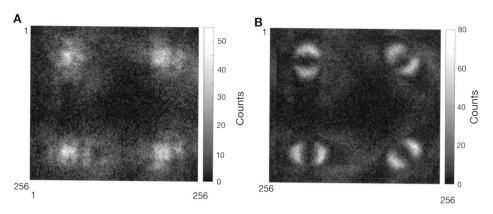

Entangled photons captured on camera at Glasgow University.

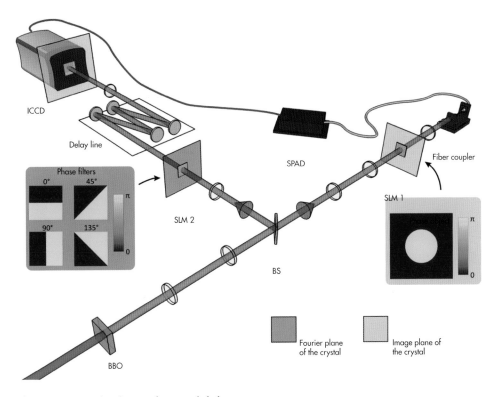

The equipment used to discover the entangled photons.

suggested that there were properties of the quantum system that remained to be discovered, which they called 'hidden variables'. Einstein accepted that quantum mechanics wasn't 'wrong'– it could accurately predict experimental outcomes – but he did argue that it wasn't complete and the EPR paradox, as it came to be known, was an attempt to demonstrate this.

One of the main tenets of quantum mechanics as we have seen is the idea of uncertainty – it isn't possible to measure all of the features of a system simultaneously, not even in theory. Another peculiar property of quantum mechanics is the phenomenon that came to be called quantum entanglement (a name coined in correspondence with Einstein by Erwin Schrödinger).

Entangled particles behave as if they are a single quantum system and not independent objects. Two photons, for example, are linked by a single wavefunction. Once they are separated, they will still share this single wavefunction, which means that measuring one will determine the state of the other. For example, the polarization of one photon will be correlated with that of the other, so determining the polarization of photon A (and therefore collapsing its wavefunction) immediately collapses the wavefunction of photon B, fixing its polarization too, even if A and B are separated by light years of space. This is known as 'non-local behaviour', but Einstein called it 'spooky action at a distance'.

Einstein and his co-authors began by setting out their premise, that if there was any way that we could learn with absolute certainty the position of a particle, and we don't disturb the particle by directly observing it, then we can say the particle exists in reality, independent of our observations. If we can take measurements of one particle that give us information about the second particle without disturbing the second particle in any way, for example measuring the momentum of the first particle gives us precise knowledge of the momentum of the second particle, it means that the second particle, which we have not directly observed, has properties that we know. It has a momentum that is real. Einstein and his collaborators argued that the assumption was being made that the process of measuring the first particle alters the reality of the second particle, instantaneously making it conform to the reality of the first particle, even if they were separated by light years of space, something they believed that 'no reasonable definition of reality could be expected to permit'. For one particle to affect the other in this way would require a faster-than-light signal to travel between them, something that was explicitly forbidden by Einstein's relativity theory.

Bohr, naturally, disagreed with Einstein's view and passionately defended the Copenhagen Interpretation of quantum mechanics. Bohr had asserted that the disturbance caused by making a measurement of a particle was what led to quantum uncertainty and had rejected Einstein's thought experiments by recourse to the uncertainty principle. If the two particles are entangled, Bohr argued, then they are effectively a single system that has a single quantum function. It still is not possible to know both the precise position and the precise momentum of the particle at the same moment. If you know the position of A then you know the position of B, and if you know the momentum of A you know the momentum of B. But it is still impossible to know both things precisely at the same moment for A, so you can't know them for B either. There is no conflict with the uncertainty principle. He also believed that the most important aspects of a quantum experiment were the conditions under which it was made. If you chose one set of conditions, for example an experiment involving wave properties, then wave properties were what you would see. If you chose something else, then you would reveal a complementary aspect to the wave properties.

None of these elements, Bohr felt, were present in the EPR thought experiment and so it failed to refute the Copenhagen Interpretation of quantum mechanics. Einstein continued to counter that quantum mechanics violated two fundamental principles: the principle of separability, which maintains that two systems separated in space have an independent existence; and the principle of locality, which says that doing something to one system cannot immediately affect the second system.

SCHRÖDINGER'S CAT

Einstein shared a thought experiment with Erwin Schrödinger that illustrated why he felt so uncomfortable with wavefunctions and probabilities. Imagine two boxes, he said, one

Schrödinger came up with his famous cat as a way of pointing out shortcomings in the Copenhagen Interpretation.

Superposition

Superposition is the idea that a quantum system can be in all its possible states at the same time until it is measured, at which point it takes up one of the basic states that form the superposition state. The system's superposition state is described by its wavefunction; the act of measurement, or observation, is said to collapse the wavefunction, causing the system to take up a definite value for the property being measured. Superposition is not the same as, for example, the case of a coin which you spin and then cover with your hand. You don't know how the coin has landed but you know that it is either 'heads' or 'tails'. In a quantum superposition state the coin would not be either heads or tails but both at the same time.

Spin is a quantum property possessed by certain particles, such as electrons. It is the spin of the electron that gives some materials their magnetic properties. Using lasers it is possible to get electrons into a superposition state where they have both up and down spin at the same time. These superposition electrons can theoretically be used in quantum computers as qubits (quantum bits) which effectively can be 'on', 'off' and something in between all at the same time. Other particles, such as polarized photons, can also be used as qubits. It was Richard Feynman who first suggested, in 1982, that enormous computing power would be unleashed if the superposition state could be exploited. Potentially qubits can be used to encode and process vastly more information than the simple binary computer bit.

contains a ball, the other is empty. Before we look in a box there is a 50 per cent chance of finding the ball. After we look, the chance of it being there is either 100 per cent or 0 per cent. But in reality, the ball was always 100 per cent in one of the boxes. The quantum mechanical view was tantamount to saying that the ball could equally well be in either box before the lid was lifted. As Niels Bohr and the Copenhagen Interpretation would have it, the ball exists in a state of superposition, actually occupying both boxes until we look and see which one it's in. The act of observation decides where it is.

Schrödinger came up with a thought experiment of his own that would pass into quantum legend. 'A cat is penned up in a box,' Schrödinger wrote, 'along with the following device: in a Geiger counter there is a tiny bit of radioactive substance, so small, that perhaps in the course of the hour one of the atoms decays, but also, with equal probability, perhaps none: if it happens... a relay releases a hammer which shatters a small flask of hydrocyanic acid.'

He explained that the wavefunction of the entire system would express the situation by having in it the living or dead cat 'mixed or smeared out'. Einstein and Schrödinger were happy that their thought experiments had demonstrated their point – there was something distinctly not right about the Copenhagen Interpretation. Einstein said that a wavefunction that 'contains the living as well as the dead cat just cannot be taken as a description of a real state of affairs.'

For Niels Bohr there was no reason why the rules of classical physics, which determine what goes on in the everyday world around us, should also apply to the quantum realm. What the quantum physicists were discovering was just the way things were, whether Einstein and Schrödinger liked it or not. At some point, an exasperated Bohr apparently declared: 'Stop telling God what to do!'

In 1964, physicist John Bell (1928–90) proposed an experiment that could test whether or not entangled particles actually did communicate with each other faster than light. Bell believed that no 'hidden variables' theory could fully explain the predictions of quantum mechanics. According to quantum theory, entangled particles remain in a superposition of states until they are measured but, as soon as one is measured, we know with certainty that the other has to have the complementary characteristic. If particle B doesn't know what happened with A it would remain in a superposition state until it too was measured. Bell assumed that each particle had determinate values, that is, ones with defined limits, and asked if such particles could reproduce the results predicted by quantum theory. He derived formulas, called the Bell inequalities, which determine how often the characteristics of particle A should correlate with those of particle B if normal probability (as opposed to quantum entanglement) was at work in determining their states. Bell proved mathematically that the predictions of quantum theory were indeed not in line with those of normal probability and that Einstein's 'hidden variables' idea just wasn't true. In the words of physicist Fritjof Capra, 'Bell's Theorem demonstrates that the universe is fundamentally interconnected'.

Experiments such as those carried out by French physicist Alain Aspect in the early 1980s, which used entangled photon pairs generated by laser, have demonstrated convincingly that 'action at a distance' is real. Aspect found that the measurements made of the entangled pairs correlated 40 times more often than would have been expected if normal probability had applied. The quantum realm is not bound by the rules of locality. When two particles are entangled, they are effectively a single system that has a single quantum function.

QUANTUM FIELD THEORY

The idea of fields carrying forces across a distance is well established in physics. A field can be thought of as anything that has values that vary across space and time. The pattern made by iron filings scattered around a bar magnet maps out the lines of force in the magnetic field, for example. Electromagnetism and other fundamental forces arise from variations in the fields that carry them. In the 1920s, quantum field theory proposed a different approach, suggesting that forces were carried by means of quantum particles, such as photons, which are the carrier particles of electromagnetism. Other particles discovered subsequently, such as the Higgs boson, the force carrier of the Higgs field which gives particles their mass, are believed to have their own associated fields.

Curved space and gravitons

Thanks to Einstein and the general theory of relativity, physicists have a workable explanation for gravitational forces as resulting from a curvature of spacetime brought about by the matter in it. In theory at least, an explanation involving an exchange of force particles, called gravitons, is equally valid; just as we can think of electromagnetism as being either the result of changes in the electromagnetic field or as an exchange of photons. The problem is that there is no current quantum theory of gravity involving gravitons that is as well worked out and proven by experiment as Einstein's relativity theory and, as yet, no experimental proof of their existence.

Quantum electrodynamics, usually referred to as QED, is the quantum field theory that deals with the electromagnetic force. The QED theory was fully developed in the late 1940s by Americans Richard Feynman, Julian Schwinger, and Japanese physicist Tomonaga Shin'ichirō, who were all working independently of each other. QED proposes that charged particles such as electrons interact with each other by emitting and absorbing photons, the force carriers of the electromagnetic force. These are 'virtual' photons, which means that they cannot be seen or detected in any way, they simply represent the force of the interaction between the charged particles, which causes them to change their speed and direction of travel as a consequence of releasing or absorbing the photon's energy. The QED theory states that the more complex the interaction, that is, the greater the number of virtual photons that are exchanged in the process, the less likely it is to occur.

QED is one of the most astonishingly accurate theories ever formulated. QED's prediction for the strength of the magnetic field associated with an electron is so close to the value produced by experiment that if the distance from London to Timbuktu was measured to the same precision it would be accurate to within literally a hairsbreadth.

Feynman diagrams

The ways in which particles can interact by the exchange of virtual photons can be visualized by means of Feynman diagrams, developed by Richard Feynman in the 1940s. Feynman diagrams have proved to be invaluable in helping scientists tackle some of the complex interactions involved in high energy physics. Each diagram represents the particles involved by wavy and straight lines and their interactions at the intersections of the lines.

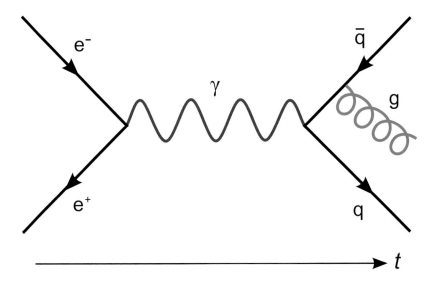

Feynman diagrams offer a visual shorthand for interactions between particles.

The success of QED provided a stepping-stone towards building quantum field theories for the other fundamental forces of nature, including the weak and strong nuclear forces. The electroweak theory holds that the electromagnetic and weak nuclear forces are actually one single force. The strong nuclear force, which binds atomic nuclei together, has its own particles, called gluons. The combined framework of forces together form the Standard Model, which underpins our understanding of particle physics and which will be explored in the next chapter.

Chapter 12

TO THE PARTICLE ZOO, AND BEYOND

To the Particle Zoo, and Beyond

TIMELINE OF PARTICLE PHYSICS

Timeline	
1858	Julius Plücker discovers the existence of cathode rays.
1895	William Röntgen discovers X-rays.
1896	Henri Becquerel discovers radioactivity.
1897	J.J. Thomson discovers the electron.
1898	Ernest Rutherford establishes that there are three types of radioactivity.
1909	Rutherford and Ernest Marsden discover the atomic nucleus. Robert Millikan calculates the size of the charge on an electron.
1911	Charles Wilson invents the cloud chamber.
1912	Victor Hess discovers cosmic rays.
1932	James Chadwick discovers the existence of the neutron. The positron is discovered by Carl Anderson.
1933	Observations by Fritz Zwicky establish the presence of large amounts of dark matter in the universe.

The idea that matter is made up of elementary particles is one that stretches back to Ancient Greece and the indivisible, indestructible atoms of Democritus (see page 61). It was an idea that was revived in the atomic theories of John Dalton and Amedeo Avogadro at the beginning of the 19th century (see page 70). Since these first proposals of its existence, the atom has gradually revealed its often surprising and unexpected nature.

The physicists of the 19th century accepted that atoms existed, but what was their nature? What *was* an atom? William Thomson, Lord Kelvin, suggested that atoms might be vortices

1939	Lise Meitner and Otto Frisch discover the process of atomic fission.
1942	Enrico Fermi produces the first controlled nuclear chain reaction.
1959	The first neutrino is detected by physicists in South Carolina.
1964	Murray Gell-Mann proposes the existence of quarks, the building blocks of matter.
1970s	The Standard Model of particle physics is developed.
1983	Physicists at CERN prove the link between the electromagnetic force and the weak force.
1984	John Schwarz and Michael Green propose string theory to unite general relativity and quantum mechanics.
2012	The Higgs Boson is discovered after an experiment at the Large Hadron Collider.

spinning in the invisible ether that was thought to pervade space. Different vortices would correspond to different chemical elements – the more complex the knot formed by the vortex the heavier the element. The theory survived for about twenty years until Michelson and Morley (see page 116) cast doubt on the existence of the ether in 1887 and so deprived the vortices of their medium.

In 1882, J.J. Thomson (1856–1940) had written an award-winning essay in which he set out a mathematical description of vortex atoms and described the ways in which they

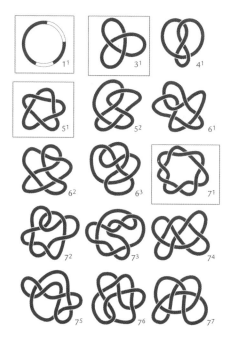

Lord Kelvin imagined atoms as being like knotted vortices.

might react chemically with each other. Some years later his researches would lead him to the discovery of one of the building blocks of atoms themselves.

DISCOVERING THE ELECTRON

In 1858, German physicist Julius Plücker (1801–68) experimented with sending a high voltage between metal plates inside a glass tube from which most of the air had been removed and discovered that a green glow was produced inside the tube near the cathode. He thought this glow was the result of rays emanating from the cathode. In 1869, Plücker's pupil Johann Hittorf, with the benefit of a more efficient vacuum, was able to discern a shadow cast by an object placed in front of the cathode, confirming that it was indeed the source of the rays and that those rays travelled out from the cathode in straight lines. English physicist and chemist William Crookes (1832–1919) carried out further investigations of these 'cathode rays' in 1879 and found that they could be bent by a magnetic field and were apparently made up of negatively charged particles which he thought were repelled from the cathode and travelled to the anode. As a result of Crookes' work on cathode rays the vacuum tubes used came to be referred to as Crookes tubes. One of the most famous examples of a Crookes tube was one in which he placed a Maltese cross, which he used to demonstrate that cathode rays travel in straight lines as its shadow could be seen at the end of the tube when a voltage was applied.

In 1883, German physicist Heinrich Hertz tried to deflect cathode rays using an electric field, but found it had no effect. As a result, Hertz concluded that cathode rays were not charged particles, but waves that could be deflected by magnetic fields, although

Julius Plücker, who first saw cathode rays.

The green glow of cathode rays inside a Crookes tube.

it wasn't clear why they should behave in this way. We now know that the reason Hertz failed to observe any deflection of the rays was that it was just too small for him to observe given the velocity of the particles and the weakness of the electrical forces involved. In 1891, Hertz also observed that the cathode rays could penetrate thin foils of gold and other metals. He saw this as analogous to the way light passed through glass, and was support for his wave theory. In 1895, however, French physicist Jean Baptiste Perrin (1870–1942) demonstrated that the cathode rays left a negative electric charge on a collector plate placed inside the cathode ray tube, offering further support for their particle nature.

J.J. Thomson began a series of experiments in 1894 that would settle the nature of the cathode rays once and for all. Working at Cambridge University's Cavendish Laboratory, he constructed a cathode ray tube with the deflector plates inside, rather than outside, the

J.J. Thomson

glass tube and discovered that cathode rays could indeed be deflected by an electric field. Thomson's experimental set up allowed him to determine the ratio of the charge of the mystery particle to its mass. Describing his work he wrote: 'I can see no escape from the conclusion that they are charges of negative electricity carried by particles of matter. The question next arises, What are these particles? Are they atoms, or molecules, or matter in a still finer state of subdivision?'

Thomson found the charge-to-mass ratio remained the same regardless of the metal used to make the electrodes, or of the composition of the gas used to fill the tube. From these observations he deduced that the particles making up the cathode ray must be something

The Millikan oil drop experiment

In 1909, Robert Millikan began a series of elegant experiments to determine the size of the charge on an electron. Oil droplets injected into an air-filled chamber pick up charges from an electric field. The drops then fall or rise under the combined influence of gravity, the viscosity of the air, and the electric field, which the experimenter can adjust. Timing the rise and fall of a drop allows the charge on it to be calculated. The value Millikan arrived at for the elementary charge, 1.592×10^{-19} coulombs, is just slightly lower than the currently accepted value of 1.602×10^{-19} C, most likely because he used an incorrect value for the viscosity of air. Millikan won the 1923 Nobel prize for his work

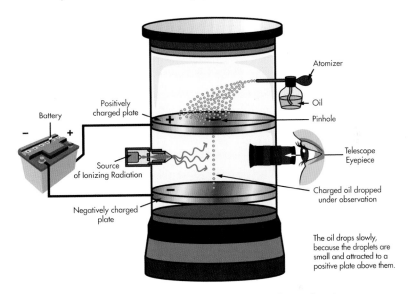

Robert Millikan's oil drop experiment allowed him to calculate the charge of an electron.

that was found in all forms of matter. By 1897 Thomson had determined that the negatively charged particles of the cathode ray had a mass that was less than 1/1000th that of a hydrogen atom. This meant that they could not be charged atoms, or indeed any other particle then known to physics. Although Thomson termed these particles 'corpuscles', the name 'electron', which had been proposed by Irish physicist George Stoney in 1891 for the fundamental unit of electrical charge, was soon adopted. Thomson was awarded the Nobel Prize in 1906 in recognition of his discovery.

The next question to be answered was how Thomson's corpuscles fitted in to the structure of the atom. Joseph Larmor, a classmate of Thomson's at Cambridge, thought that they weren't part of the atom at all, but rather a constituent of the invisible ether. It was known that atoms

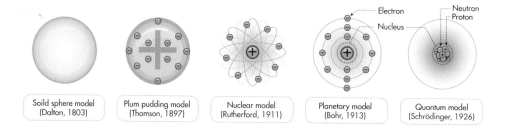

Soild sphere model
(Dalton, 1803)

Plum pudding model
(Thomson, 1897)

Nuclear model
(Rutherford, 1911)

Planetary model
(Bohr, 1913)

Quantum model
(Schrödinger, 1926)

Science's conception of the atom has changed greatly over the years.

were electrically neutral, so in order to balance the negative charge of the electrons Thomson himself proposed that they were embedded in a positively charged cloud, like raisins in a cake, an image that gave rise to Thomson's atom being dubbed the Plum Pudding model. Thomson's model was important in that, for the first time, it set out a description of the atom as something divisible. Within a few years, however, the Plum Pudding was running into problems.

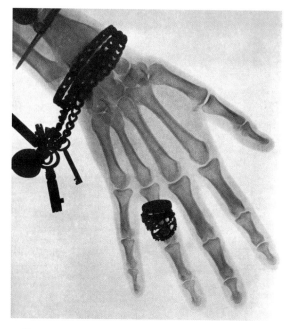

Wilhelm Röntgen used his newly discovered x-rays to capture this groundbreaking image of his wife's hand.

RADIOACTIVITY

In 1896, around the same time as Thomson was carrying out his experiments, French physicist Henri Becquerel (1852–1908) was making discoveries of his own. Becquerel was studying the properties of x-rays, which had been discovered the previous year by Wilhelm Röntgen (1845–1923). Röntgen was working with a cathode-ray tube in his laboratory, when he noticed that a nearby fluorescent screen had started to glow. He concluded that a new type of ray was being emitted from the tube and later experiments established that this x-ray could pass through most substances, including through the soft tissue of humans, but not bones and metal objects. This discovery would win Röntgen the first Nobel Prize for Physics in 1901.

Becquerel believed that uranium absorbed the sun's energy and then emitted it as x-rays. He exposed a uranium-containing compound to sunlight and then placed it on photographic plates wrapped in black paper. His experiment was frustrated because the day was overcast, but Becquerel decided to develop his photographic plates anyway. To his surprise, the outlines left by the compound were strong and clear, proving that the uranium emitted radiation without an external source of energy such as the sun. He found that the uranium compound continued to emit energy which appeared not to diminish over time, even over a period of several months, and that pure metallic uranium worked even better.

Becquerel had discovered radioactivity, a word that was coined by Marie Curie (1867–1934) in 1898. Further investigations by Becquerel

Henri Becquerel, the discoverer of radioactivity.

and others, such as Marie Curie and her husband Pierre (1859–1906), revealed that other substances had radioactive properties as well. The Curies discovered that samples of pitchblende, a mineral that contains uranium, appeared to produce more radioactivity than pure uranium. They surmised that there had to be another radioactive substance present. Eventually they isolated a sample of a new chemical element, over 300 times more radioactive than uranium, which they called polonium. But that wasn't the end of the story. The waste left behind after the polonium had been extracted was still highly radioactive.

After some years of arduous work that involved grinding, filtering and dissolving 20-kilogram samples of pitchblende from which the uranium had already been extracted, in 1902 Marie Curie succeeded in isolating a small amount of the element the Curies called radium. In 1903, Marie and Pierre Curie were awarded the Nobel Prize for Physics jointly

with Henri Becquerel for their work on radioactivity.

In 1898, using a simple experimental set up, New Zealand-born Ernest Rutherford, who had worked in the Cavendish Laboratory with J.J. Thomson, discovered that there were three different types of radioactivity. Based at McGill University in Montreal, Rutherford took a sample of uranium as his source of radioactivity and an electroscope detector, and placed increasing thicknesses of aluminium foil between them, first one, then two, and so on up to thirteen. At each stage, he measured the intensity of the radiation by measuring the time required to discharge the electroscope. He found at least two distinct types of radiation which, simply for convenience, he designated α (alpha) and β (beta). He established that the α and β rays were, respectively,

Marie Curie's life was almost certainly shortened by her exposure to the then unknown dangers of radioactivity.

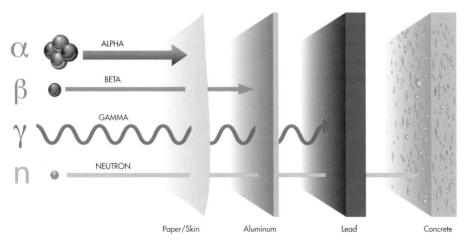

Different types of radiation.

positively charged and negatively charged particles. In 1901 he determined that Becquerel's rays were electromagnetic in nature and dubbed them γ (gamma) rays. In that same year Rutherford and chemist Frederick Soddy found that one radioactive element can decay into another, a discovery that earned Rutherford the 1908 Nobel Prize in Chemistry, an award that irked him as he considered himself a physicist, not a chemist.

INTO THE NUCLEUS

In 1907 Rutherford took up a post at the University of Manchester, England. In 1909, he set student Ernest Marsden the task of carrying out an experiment that involved firing alpha particles from a radioactive source at a thin gold foil. Any scattered particles would hit a screen coated with zinc sulphide, which sparkles when struck by charged particles. Marsden, who had to sit in a darkened room patiently staring at the screen, was expected to see nothing at all, but instead he saw fleeting flashes, on average around one per second.

He reported the results to an astonished Rutherford. According to Thomson's plum pudding model, the positive charge of the 'pudding' should be widely distributed throughout

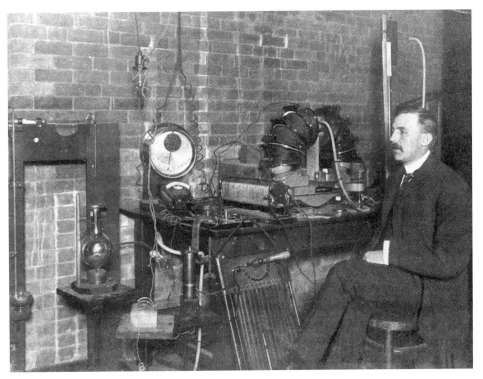

Ernest Rutherford was responsible for several breakthrough discoveries, including that one radioactive element could decay into another.

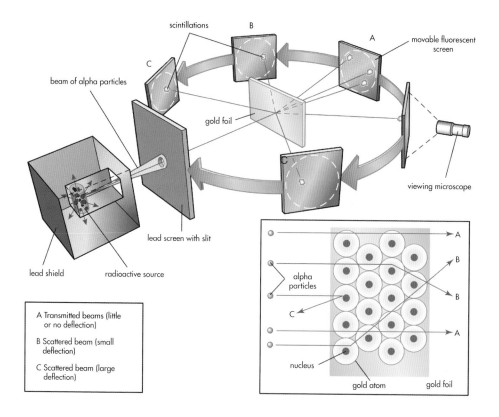

scintillations
B
A movable fluorescent screen
C
beam of alpha particles
gold foil
viewing microscope
lead screen with slit
lead shield   radioactive source

A Transmitted beams (little
  or no deflection)

B Scattered beam (small
  deflection)

C Scattered beam (large
  deflection)

alpha particles
A
B
B
C
A
nucleus
gold atom   gold foil

A diagrammatic representation of Rutherford's 'gold foil' experiment, which revealed the existence of the nucleus when some alpha particles bounced right back.

the volume of the atom. The large, fast-moving alpha particles should have passed through the positive pudding in the gold foil with scarcely any deflection as the electric field in the atom was too weak to affect them. As Rutherford later remarked, 'It was as if you fired a 15-inch shell at a piece of tissue paper and it came back and hit you.'

Rutherford concluded that the alpha particles must have been repelled by large positively charged particles lurking within the atoms. This led Rutherford to propose a new model of the atom in 1911, in which most of the mass of the positive charge is concentrated in a nucleus situated in the middle of the atom, with the electrons orbiting this central core, like planets around a star. Rutherford calculated the size of the nucleus and found it to be only about $\frac{1}{100,000}$th the size of the atom. The atom, it appeared, was mostly empty space.

Rutherford's model of the atom was not immediately accepted by his fellow physicists. According to James Clerk Maxwell's equations, electrons travelling in a curved path should radiate electromagnetic energy, eventually slowing down and falling into the nucleus. A solar system atom would be short lived. Fortunately, Niels Bohr and the new ideas from quantum mechanics came to the rescue of Rutherford's model. Bohr showed that the atom could persist if electrons were only allowed to occupy certain discrete orbitals of fixed size and energy. Radiation only occurs when an electron jumps from one orbit to another. The atom will be completely stable when it is in the state with the smallest orbit, since there is no orbit of lower energy into which the electron can jump. This revised Rutherford–Bohr model of the atom, or simply the Bohr model, was presented in 1913. In 1926, Erwin Schrödinger proposed that, rather than the electrons moving in fixed orbits or shells, they behaved as waves (see page 150). Schrödinger produced a model for the distributions of electrons in an atom that showed the nucleus surrounded by clouds of electron density. We can't know *exactly* where the electrons are, just where they are most likely to be. These regions of probability are the electron orbitals.

PROTONS AND NEUTRONS

Between 1914 and 1919 Rutherford carried on his experiments in Manchester. One of these involved bombarding nitrogen gas with alpha particles. Rutherford hypothesized that the radiation being emitted as a result of the bombardment might be the nucleus of a hydrogen atom. In 1919 Rutherford succeeded J.J. Thomson as head of the Cavendish Laboratory in Cambridge. At Rutherford's suggestion Patrick Blackett carried out further research in the

The Cloud Chamber

Trails left by alpha particles in a cloud chamber.

The first cloud chamber was built by physicist Charles Wilson (1869–1959) in 1911. He was inspired by the morning mist rising from Scottish hilltops. Air inside the chamber was saturated with water vapour and then the pressure was lowered. When, for example, a positively charged alpha particle moved through the chamber it removed electrons from the gas, leaving charged atoms which attracted the water vapour and caused a visible trail to form.

1920s, capturing cloud chamber images that revealed that some of the alpha particles were being absorbed by the nitrogen nuclei resulting in an oxygen atom being formed and a hydrogen nucleus being emitted. At the heart of the atom, Rutherford believed, was a particle with a positive charge, which he named the proton, from the Greek word 'protos', meaning 'first'. Atoms of different elements have different numbers of protons, with the hydrogen nucleus, the smallest atom, having a single proton.

As the studies of atomic disintegration proceeded it became apparent that the atomic number of the atom (the number of protons in the nucleus, equivalent to the positive charge of the atom) was less than the atom's atomic mass. A helium atom, for instance, has an atomic mass of 4, but an atomic number (or positive charge) of 2. Since J.J. Thomson had established that electrons weren't massive enough to account for the difference, it seemed that there must be something besides the protons in the nucleus.

One suggestion was that electrons, along with additional protons, were present in the nucleus, but the negatively charged electrons cancelled out the positive charge on the protons, meaning that the protons still contributed their mass but not their charge. A helium nucleus, therefore, contained four protons and two electrons yielding a mass of 4 but a charge of only 2. Rutherford also suggested that there could be a particle, consisting of a paired proton and electron, which he called a neutron, that had a mass similar to that of the proton but no charge.

James Chadwick, the discoverer of the neutron.

An assistant director at the Cavendish, and former PhD student of Rutherford's, was James Chadwick (1891–1974), who was himself researching radioactivity. He was aware of experiments being carried out in Europe by Frédéric and Irène Joliot-Curie who had been studying the particle radiation emitted by beryllium. The Joliot-Curies believed the radiation took the form of high-energy photons, but Chadwick was unconvinced by this. Massless photons wouldn't knock loose particles as heavy as protons, as had been observed to happen.

In 1932, Chadwick tried out the experiments for himself and came to the conclusion that the beryllium radiation was a neutral particle similar in mass to the proton. He succeeded in demonstrating that the neutron did indeed exist and that its mass was about 0.1 per cent more than that of the proton. He published his findings in a paper entitled 'The Possible Existence of a Neutron' and in 1935 received the Nobel Prize for his discovery.

Werner Heisenberg showed that the neutron could not be a proton-electron pairing as Rutherford had suggested but was a unique fundamental particle in its own right. The discovery changed physicists' conception of the atom and they soon found that the chargeless, massive neutron made an ideal projectile for bombarding atomic nuclei, because, unlike positively

Atomic energy

In 1939, physicists Lise Meitner and Otto Frisch showed how the nucleus of the uranium atom could split into two parts, the process of atomic fission, producing free neutrons, which could go on to split further uranium atoms, and a huge release of energy. In 1942, Enrico Fermi and his team, working in a squash court at the University of Chicago, built on this work to produce the first controlled nuclear chain reaction. Fermi used metal rods to absorb the neutrons released, allowing him to control the rate of the reaction. Within three years the Manhattan Project team triggered a device that released its energy in a devastating uncontrolled runaway reaction when the first atomic bomb was detonated in the desert of New Mexico.

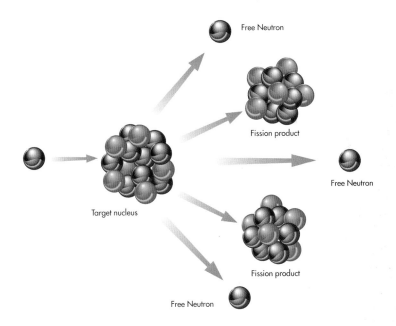

Nuclear fission occurs when the nucleus of an atom splits apart.

charged alpha particles, it was not repelled by the similarly charged nucleus and could smash right into it. Before long, neutron bombardment of the uranium atom was being used to split its nucleus and release the huge amounts of energy predicted by Einstein's $E=mc^2$ and open up the possibility of the atomic bomb. Chadwick himself was one of the scientists who worked on the Manhattan Project to develop the bomb during World War II.

COSMIC EXPLORERS

In August 1912, Austrian physicist Victor Hess ascended to 5,300 metres in a balloon. Measuring the rate of ionization in the high atmosphere he found that it was some three times that at sea level. He had discovered cosmic rays (a term coined by Robert Millikan in 1926) – high-energy particles entering the atmosphere from outer space.

Cosmic rays opened up a hitherto unknown world of subatomic particles. In 1932, Carl Anderson was studying cosmic particles in a cloud chamber at the California Institute of Technology (Caltech) when he spotted something with the same mass as an electron, but positively charged. Further observations led him to conclude the tracks were actually due to antielectrons, produced alongside an electron from the impact of cosmic rays in the cloud chamber. He called the antielectron a 'positron'. It was vindication of a prediction of the existence of antiparticles made by Paul Dirac a few years earlier (see page 151).

Victor Hess, who detected cosmic rays in the upper atmosphere.

In 1936, Anderson played a part in the discovery of an altogether new particle. While studying cosmic radiation cloud chamber trails with Seth Nedermeyer at Caltech, Anderson noticed signs of negatively charged particles that curved more sharply than electrons, but less than protons when passing through a magnetic field. The conclusion was drawn that the new particle must be somewhere between the electron and proton in mass. The mesotron, as Anderson called it, was at first thought to be a particle that had been predicted to exist by Japanese

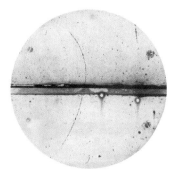

A cloud chamber image that revealed the existence of the positron.

Yukawa Hideki postulated a force carrier that bound protons and neutrons together in the nucleus.

theoretical physicist Yukawa Hideki in 1935 to explain the force that binds protons and neutrons together in the atomic nucleus but it was discovered to have the wrong properties. Yukawa's particle, the pi meson, was eventually discovered in 1947 and the mesotron was renamed the mu meson, adopting the more general term meson to refer to any particle with a mass intermediate between that of electrons and protons and neutrons. As further experiments revealed the existence of yet more mesons it was discovered that the mu meson didn't share the same properties as the other mesons and it was renamed again as the muon.

TOWARDS THE STANDARD MODEL

By 1932, physicists had established that atoms were formed from three particles – electrons, protons and neutrons. Along with the photon described by Planck and Einstein, it was a simple and satisfying set up but there were fundamental problems that had to be solved. Not least of these was what held the nucleus together. Why didn't the positively charged protons in the nucleus push each other away? Soon things became yet more complicated by the discovery of more and more subatomic particles – by the 1960s there were hundreds of them. Physicists began sorting this burgeoning particle zoo into more manageable categories. There were the hadrons, which included the baryons, heavier particles such as the proton and neutron and their corresponding antiparticles, as well as the intermediate mesons; and the lighter leptons, such as the electron and the theoretically predicted but as yet unobserved neutrino. How could it all be brought together in a coherent whole?

Discovering the neutrino

The existence of the neutrino was first predicted in 1931 when Wolfgang Pauli theorized the existence of an unknown particle to account for an apparent loss of energy and momentum he detected in his studies of radioactive decay. The elusive particle was first detected in 1959 by physicists Frederick Reines and Clyde Cowan in the course of experiments at a nuclear reactor in South Carolina. We now know that neutrinos are everywhere, permeating everything in the universe. Every second over 100 billion neutrinos pass through every square inch of your body. They barely interact with any other particles and can pass unperturbed right through the earth. There are three types of neutrinos: the electron neutrino, muon neutrino and tau neutrino (the electron neutrino was the first to be discovered).

QUARKS AND THE EIGHTFOLD WAY

In 1962, Murray Gell-Mann and Yuval Ne'eman proposed a possible solution to the proliferation of particles perplexing physicists. They devised a scheme for classifying the hadrons called the Eightfold Way, based on a mathematical symmetry known as SU(3) and named after Buddha's Eightfold Path to enlightenment.

The Eightfold Way, rather like the Periodic Table of the elements used by chemists, could be used to describe and categorize the characteristics of particles that had been discovered, such as the magnetic properties of protons, and predict the properties of particles that hadn't even been seen yet. Underpinning all of this, Gell-Mann proposed in 1964, were three new elementary particles, which he named quarks. (Georg Zweig had also come up with this idea in the same year, calling his particles 'aces'.) Experiments at the Stanford Linear Accelerator in the late 1960s confirmed the existence of quarks.

Quarks, along with leptons, appear to be the true building blocks of matter. There are believed to be six 'flavours' of quark – up, down, charm, strange, top and bottom – combinations

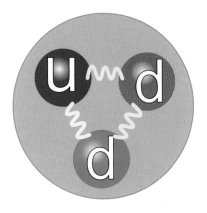
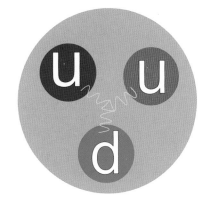

Quark structure of a neutron (left) and a proton (right).

of which can successfully account for the more than 200 types of meson and baryon known to physics. Quarks have fractional electric charge, the up quark, for example, has a charge of $+\frac{2}{3}$ and the down quark a charge of $-\frac{1}{3}$. The familiar protons and neutrons are constructed from three up and down quarks (up-up-down – total charge +1 – and up-down-down – total charge 0 – respectively). The force binding quarks together, called the colour force, is so powerful that quarks have never been detected in isolation. The theory that describes the interactions between quarks is called quantum chromodynamics.

THE STANDARD MODEL

The Standard Model, developed in the early 1970s, is a mathematical model that aims to bring together all that we know about particles and forces into a coherent whole. It postulates that everything in the universe is formed from a few basic building blocks – 31 in total – called fundamental particles, and the interactions between them are governed by four fundamental forces.

The fundamental forces work over different ranges and have different strengths. Gravity is the weakest, but its range is infinite. The electromagnetic force also has infinite range but is many times stronger than gravity. The weak and strong forces are effective only at the level of subatomic particles; the weak force, despite its name, is much stronger than gravity but weaker than the electromagnetic force. The strong force, as the name suggests, is the strongest of the four.

In the 1940s a theory had emerged that envisaged the electromagnetic force in terms of quantum fields, and electrons and photons arising from fluctuations in these fields. This was quantum electrodynamics (QED) (see page 159) and its power to explain how light interacts with matter and the workings of the electromagnetic force soon became evident in a series

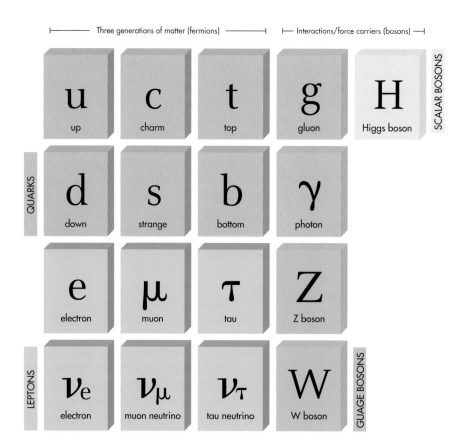

The fundamental particles of the Standard Model.

of experiments. QED's ideas were extended to fields for other fundamental forces: the strong nuclear force (which binds the atom's nucleus together and stops the protons flying apart) and the weak nuclear force (which describes how atoms decay and emit radiation). The strong nuclear force is considered to be a residual effect of the quark colour force, extending beyond the boundary of the proton or neutron. These forces result from matter particles exchanging force-mediating particles. As the electromagnetic field has the photon, so the colour force and the strong nuclear force have the gluon and the weak force has W and Z particles. The force-carrier particles belong to a group of particles called 'bosons'. Particles of matter transfer energy by exchanging bosons with each other.

Although it has never been detected, the 'graviton' is the theoretical force-carrying particle of gravity (see page 159). At the scale of the atom the effects of gravity are so negligible they can effectively be discounted. It is only at the macro scale of humans and galaxies that gravity dominates. The quantum rules of the atomic world, and the general theory of relativity that guides the macro world, resist fitting into a single mathematically compatible whole within the context of the Standard Model. The Standard Model holds up well, but its biggest shortcoming remains its failure to accommodate the force with which we are most familiar: gravity.

HUNTING FOR HIGGS

The W and Z bosons responsible for carrying the weak force were predicted by Nobel laureates Steven Weinberg, Abdus Salam and Sheldon Glashow in the 1960s. It is the weak force that triggers nuclear fusion and causes stars to shine. In 1963, Glashow, Salam and Weinberg suggested that the weak nuclear force and the electromagnetic force could be combined in what would be called the electroweak force. They predicted that this would occur at energy and temperature levels similar to those which were found shortly after the Big Bang, when the universe expanded rapidly from a super-dense subatomic state. In 1983, physicists

The Compact Muon Solenoid at the CERN nuclear laboratory which was involved in detecting the Higgs boson.

at CERN, the European Laboratory for Particle Physics, achieved these temperatures in a particle accelerator and showed that the electromagnetic force and weak nuclear force were indeed related.

However, the equations that described the unified force required the force-carrying particles to be massless. While this was certainly true of the photon, scientists knew that the W and Z bosons would have to be heavy to account for their short range, nearly 100 times more massive than the proton in fact. Theorists Robert Brout, François Englert and Peter Higgs proposed a solution to the problem by suggesting that what is now called the Brout-Englert-Higgs mechanism gives mass to the W and Z particles through their interaction with the 'Higgs field', which is believed to permeate the entire universe.

The more a particle interacts with the Higgs field the more mass it has, and the heavier it is. As the universe cooled and expanded following the Big Bang the Higgs field grew with it. Particles like the photon do not interact with it and have no mass at all. In common with all fundamental fields, the Higgs field has an associated force-carrying particle – the Higgs boson. But could it be found?

On 4 July 2012, scientists at CERN's Large Hadron Collider announced they had observed a new particle of a mass that was consistent with the predicted value for the Higgs boson. Two teams worked separately, not discussing their results, to ensure that the findings were genuine. It will still take many years of study to be absolutely certain that the Higgs boson has indeed been discovered but as of 2019 all of the findings from CERN continue to confirm and validate the earlier observations.

On 8 October 2013 the Nobel prize in physics was awarded jointly to François Englert and Peter Higgs 'for the theoretical discovery of a mechanism that contributes to our understanding of the origin of mass of subatomic particles, and which recently was confirmed through the discovery of the predicted fundamental particle, by the ATLAS and CMS experiments at CERN's Large Hadron Collider'.

WHERE IS ALL THE ANTIMATTER?

Since the discovery in 1932 of the positron, or antielectron, it has been confirmed that all matter has an antimatter equivalent. Antimatter particles have the same mass as their matter counterparts but are opposite in characteristics such as electric charge. Matter and antimatter particles are always produced as a pair and if they come into contact with each other both are converted to energy in the form of photons and annihilated. According to theory, equal amounts of matter and antimatter should have been created in the Big Bang, yet somehow enough matter survived so that today everything we see is composed almost entirely of matter. Where did all the antimatter go? One of physics' greatest challenges is to explain the asymmetry between matter and antimatter in the universe. A solution involved violating what was then thought to be a fundamental symmetry in nature.

The experiment at Brookhaven National Laboratory that discovered evidence for CP violation.

CP symmetry has two components: charge conjugation (C) and parity (P). Charge conjugation transforms a particle into its corresponding antiparticle, mapping matter into antimatter. According to charge conjugation symmetry the laws of physics apply equally to particle and antiparticle. Parity reverses the space coordinates. Applying P to an electron moving with a velocity v from left to right, flips its direction so it is now moving with a velocity $-v$, from right to left. Parity conservation means that the mirror images of a reaction occur at the same rate – if particles are being emitted up and to the right an equal number should be emitted down and to the left. Applying CP to matter gives us the corresponding antimatter mirror image.

In 1964, physicists James Cronin and Val Fitch made the surprising discovery that particles called neutral kaons, which were formed from a strange quark and a down antiquark did not obey the CP symmetry rules. The neutral kaons could transform into their antiparticles (in which each quark is replaced by its opposite) but with different probabilities of each happening. The difference was small, just one in a thousand, but it was enough to demonstrate a difference between matter and antimatter. Symmetry had been violated. Fitch and Cronin's discovery won them the 1980 Nobel Prize.

In 1972, Japanese theoretical physicists Kobayashi Makoto and Maskawa Toshihide brought CP violation into the Standard Model by proposing the existence of six types of quark. With six quarks, quantum mixing would allow for the occurrence of very rare CP-violating decays. Their predictions were borne out by the discovery of the bottom and top quarks, in 1977 and 1995, respectively. Unfortunately, this theory is still unable to provide a full explanation for the amount of matter in the universe as its predictions are still several orders of magnitude short of what is observed. There is evidently some as yet hidden process at work that tips the odds in favour of matter over antimatter, but just what that process might be remains elusive.

STRINGING TOGETHER A THEORY OF EVERYTHING

Einstein's relativity theories provide a framework for understanding the universe on the grand scale of stars and galaxies; quantum theory describes how it works on the atomic scale. Both theories have been tried and tested to unimaginable levels of accuracy, but the problem remains that there seems no way to unite the two. Einstein himself spent the last 30 years of his life trying to unite gravity and electromagnetism – and failed. We have seen

An artistic impression of vibrating superstrings.

that electromagnetism can be united with the weak nuclear force in the electroweak force. Today, physicists believe that in the energy-dense early universe, shortly after the Big Bang, the strong nuclear force was united with the other two and speculate that for a vanishingly brief split-second gravity might have been included too.

Currently, one of the most hopeful candidates for a theory of everything is string theory. It encompasses a theory of gravity on the microscopic scale, provides a unified and consistent description of the fundamental structure of the universe, and unites all four fundamental forces and the fundamental particles of the Standard Model. In December 1984, John Schwarz, of the California Institute of Technology, and Michael Green, of Queen Mary College, London, published a paper showing that string theory could build a bridge across the yawning mathematical chasm that separates general relativity and quantum mechanics.

Wake up little SUSY

The theory of supersymmetry (SUSY), championed by physicists such as Japanese-American Bunji Sakita and Italian Bruno Zumino, postulates that every member of the Standard Model's particle zoo has a heavier twin. Quarks, for example, are partnered by 'squarks' (supersymmetric quarks). If supersymmetry theory is correct, then these super-twins may be the source of dark matter (see below). SUSY is an important feature of string theory.

At the heart of string theory is the idea that all of the fundamental particles of the Standard Model are really just different aspects of one basic object: a string. Nature's fundamental particles, its electrons, quarks and neutrinos, have thus far been portrayed as being as-small-as-you-can-go objects with no internal structure. String theory challenges this. It proposes that at the heart of every particle is a tiny, vibrating string-like filament. The differences between one particle and another – their mass, charge and other properties – all depend on the vibrations of their internal strings. Like a skilled violinist conjuring a mesmerizing melody, nature manifests all the particles of the atomic realm through changes in the frequency of a one-dimensional subatomic string. It is particularly intriguing that one of the 'notes' produced by the string corresponds to the graviton, the hypothetical particle that carries the force of gravity from one location to another, just as the photon does for the electromagnetic force.

Is there any evidence that strings have a basis in reality? The mathematics of string theory require them to be about a million billion times smaller than anything the world's most powerful particle accelerators have uncovered. According to physicist Brian Greene, it would require 'a collider the size of the galaxy', to pull these strings into view.

An additional complication of string theory is that its equations require extra dimensions of space to make them work. In 1918, Hermann Weyl had proposed a scheme based on a generalization of the geometry of curved space developed in Einstein's general theory of relativity. Theodor Kaluza had gone on to show that if spacetime were extended to five dimensions, then four of those dimensions would encompass Einstein's general relativity equations, while the fifth dimension would be the equivalent for Maxwell's equations for electromagnetism. Oskar Klein later determined that the fifth dimension would be curled up so small that we can't detect it. String theorists took up these ideas and suggested that the universe has the three big dimensions that we all know and move through, but that there might also be additional incredibly tiny and compact dimensions, wound up inside the 'normal' three, that were beyond our ability to detect.

A group of theorists suggested that because strings are so small they will vibrate not only in the 'big' dimensions, but also in the tiny ones. Audaciously, they predicted that since the properties of the fundamental particles, which we can detect experimentally, are determined by the vibrations of the strings, and the vibrations are in turn determined by the shape of the extra dimensions, there might be a way to work back to a map of the esoteric dimensions.

Everything we know about the Universe is based around four fundamental properties – space, time, mass and energy. Einstein's explanation for gravity is that it is a result of spacetime being curved by the presence of mass and energy. In 1971, with his proposal for Hawking radiation, by means of which black holes slowly evaporate, Stephen Hawking showed that sharply curved spacetime could give rise to mass-energy (according to Einstein mass and energy are just two different forms of the same thing).

So if mass-energy curves spacetime and curved spacetime causes mass-energy to appear – which came first? If they both arose together simultaneously in the Big Bang then what did they arise from? What in the Universe could be more basic than space, time, mass and energy? No one knows the answers to this question or even if the question itself is actually meaningful in any way that we could understand.

MAXIMUM DARKNESS

As a consequence of the Big Bang the universe is expanding. An obvious question to ask is whether this expansion will continue forever, or will it eventually slow down and even reverse? Theoretically, at some point gravity should begin to slow the rate of expansion. Or so it was thought. To the surprise of scientists, observations from the Hubble Space Telescope revealed that the universe is actually expanding faster now than it was in the distant past. This was totally unexpected and without any obvious explanation. The solution remains a puzzle, but it does have a name – dark energy.

At present, physicists have no idea what dark energy actually is, but they do have a good idea how much there is. Based on the accelerating rate of expansion, roughly 68 per cent

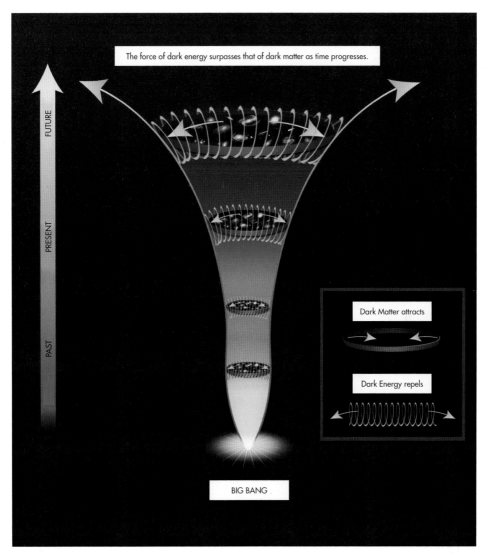

The force of dark energy surpasses that of dark matter as time progresses.

FUTURE

PRESENT

PAST

Dark Matter attracts

Dark Energy repels

BIG BANG

The tendency of dark energy to push the universe apart outweighs dark matter's tendency to pull it together.

(more than two thirds) of the universe is composed of dark energy. One explanation is that dark energy is a property of space. There is no space waiting there for the universe to expand into, rather space is created as the universe expands and energy inherent in the fabric of space causes the universe to expand at an ever-faster rate in a self-perpetuating process. One explanation for how space acquires energy comes from quantum theory, which allows for temporary 'virtual' particles to pop in and out of existence. Unfortunately, when physicists crunched the numbers to calculate how much energy this would provide the answer came

The blue regions in this image of colliding galaxy clusters show where invisible dark matter is located only detectable by its gravitational influence.

out wrong by a factor of 10^{120}. Another possibility, equally hard to accept, is that Einstein got it wrong, that general relativity is fundamentally flawed in some way, and we need a new theory of gravity that gives a better fit for the expanding universe. Perhaps one day someone who will be for Einstein what he was for Newton will step forward, but as yet there are no obvious candidates.

On top of the mystery of dark energy there is a further cosmological conundrum – dark matter. In 1933, astronomer Fritz Zwicky's studies of the motions of galaxies suggested that a significant amount of the mass of galaxies was undetectable. Further observations have led to the conclusion that 27 per cent of the universe is made up of dark matter. Add that to the 68 per cent that is composed of dark energy and we are left with just 5 per cent for everything else. And that everything else is all that we can see and know. All of the ideas and theories and discoveries in this book apply to just one-twentieth of the universe.

The rest is shrouded in darkness.

INDEX

PICTURE CREDITS

t = top, b = bottom

Alamy: 17, 49, 52, 63, 77t

Bridgeman Images: 42, 54b, 55, 71

Caltech: 138 (MIT/LIGO)

Getty Images: 19, 20, 25t, 27, 30, 35, 53t, 86, 94, 97, 108, 113, 118b, 146b, 148, 166, 168b, 172, 177

Library of Congress: 72

NASA: 136, 187, 188

Public Domain: 9, 11, 12t, 12b, 14, 18, 21, 22, 24t, 24b, 25b, 26t, 31b, 32, 34, 37, 38, 41, 46, 47, 48, 51, 53b, 54t, 56, 57b, 60t, 60b, 64, 69, 70, 75b, 78, 79t, 81, 89, 96, 99, 102, 103, 106, 107, 109, 110, 112, 115, 119, 120, 121, 122, 123, 127, 128, 132, 134, 135, 146t, 151, 153, 157, 160, 164t, 164b, 165, 169, 170t, 171, 174, 176b, 179, 180, 181

Science Photo Library: 31t, 58, 59, 77b, 79b, 87, 92, 93, 100, 111, 117, 118t, 133, 150, 173, 176t, 183, 184

Shutterstock: 7, 8, 15, 26b, 33, 39, 45, 73, 75t, 125, 131, 137, 139, 143, 145, 147, 149, 161, 167, 168t, 170b, 175

University of Glasgow: 154, 155

Wellcome Collection: 10, 13, 29, 40, 43, 57t, 62t, 62b, 66, 67, 76, 82, 83, 85, 90, 91, 101, 105, 142, 144